Vishal Santosh Sharma · Uday Shanker Dixit ·
Ajay Gupta · Rajeev Verma · Varun Sharma
Editors

Machining and Additive Manufacturing

Select Proceedings of CPIE 2023

Springer

Editors
Vishal Santosh Sharma
School of Mechanical, Industrial
and Aeronautical Engineering
University of the Witwatersrand
Johannesburg, South Africa

Ajay Gupta
Department of Industrial and Production
Engineering
Dr. B. R. Ambedkar National Institute
of Technology
Jalandhar, Punjab, India

Varun Sharma
Department of Industrial and Production
Engineering
Dr. B. R. Ambedkar National Institute
of Technology
Jalandhar, Punjab, India

Uday Shanker Dixit
Department of Mechanical Engineering
Indian Institute of Technology Guwahati
North Guwahati, Assam, India

Rajeev Verma
Department of Industrial and Production
Engineering
Dr. B. R. Ambedkar National Institute
of Technology
Jalandhar, Punjab, India

ISSN 2195-4356 ISSN 2195-4364 (electronic)
Lecture Notes in Mechanical Engineering
ISBN 978-981-99-6093-4 ISBN 978-981-99-6094-1 (eBook)
https://doi.org/10.1007/978-981-99-6094-1

This Springer imprint is published by the registered company Springer Nature Singapore Pte Ltd.
The registered company address is: 152 Beach Road, #21-01/04 Gateway East, Singapore 189721, Singapore

Lecture Notes in Mechanical Engineering

Lecture Notes in Mechanical Engineering (LNME) publishes the latest developments in Mechanical Engineering—quickly, informally and with high quality. Original research reported in proceedings and post-proceedings represents the core of LNME. Volumes published in LNME embrace all aspects, subfields and new challenges of mechanical engineering.

To submit a proposal or request further information, please contact the Springer Editor of your location:

Europe, USA, Africa: Leontina Di Cecco at Leontina.dicecco@springer.com
China: Ella Zhang at ella.zhang@springer.com
India: Priya Vyas at priya.vyas@springer.com
Rest of Asia, Australia, New Zealand: Swati Meherishi at swati.meherishi@springer.com

Topics in the series include:

- Engineering Design
- Machinery and Machine Elements
- Mechanical Structures and Stress Analysis
- Automotive Engineering
- Engine Technology
- Aerospace Technology and Astronautics
- Nanotechnology and Microengineering
- Control, Robotics, Mechatronics
- MEMS
- Theoretical and Applied Mechanics
- Dynamical Systems, Control
- Fluid Mechanics
- Engineering Thermodynamics, Heat and Mass Transfer
- Manufacturing
- Precision Engineering, Instrumentation, Measurement
- Materials Engineering
- Tribology and Surface Technology

Indexed by SCOPUS, EI Compendex, and INSPEC.

All books published in the series are evaluated by Web of Science for the Conference Proceedings Citation Index (CPCI).

To submit a proposal for a monograph, please check our Springer Tracts in Mechanical Engineering at https://link.springer.com/bookseries/11693.

Preface

It gives immense pleasure and satisfaction to present the select proceedings from CPIE-2023, titled *Machining and Additive Manufacturing*, before the readers. This volume brings together a diverse range of research papers presented at the conference, highlighting the latest advancements and innovations in the field of manufacturing.

Machining and Additive Manufacturing has always been at the forefront of technological advancements, shaping industries and driving progress in various sectors. This collection aims to provide a comprehensive overview of the current trends, challenges, and solutions in these areas, presenting a valuable resource for researchers, engineers, and practitioners alike.

The papers included in this volume cover a wide range of topics, encompassing both traditional machining techniques and cutting-edge additive manufacturing technologies. From the fundamentals of machining processes to the exploration of novel materials and the optimization of manufacturing systems, the contributions offer a wealth of insights into the ever-evolving landscape of manufacturing.

Furthermore, this collection showcases the interdisciplinary nature of machining and additive manufacturing, highlighting the integration of engineering principles, materials science, computer-aided design, and advanced manufacturing techniques. The research presented here explores the symbiotic relationship between these domains, fostering collaboration and cross-pollination of ideas to drive innovation forward.

We would like to express our gratitude to all the authors who have contributed their valuable research to this volume. Their expertise, dedication, and intellectual curiosity have made this collection a testament to the remarkable progress being made in the field of machining and additive manufacturing.

We also extend our thanks to the reviewers and editors who have diligently worked to ensure the quality and rigor of the papers included in this volume. Their expertise and commitment to scholarly excellence have played a vital role in shaping this collection.

We hope that the research presented in *Machining and Additive Manufacturing* will inspire further exploration, spark new collaborations, and contribute to the advancement of manufacturing technologies. May this volume serve as a stepping stone toward a future where machining and additive manufacturing continue to revolutionize industries, drive innovation, and shape the world we live in.

Johannesburg, South Africa — Prof. Vishal Santosh Sharma
North Guwahati, India — Prof. Dr. Uday Shanker Dixit
Jalandhar, Punjab, India — Dr. Ajay Gupta
Jalandhar, Punjab, India — Dr. Rajeev Verma
Jalandhar, Punjab, India — Dr. Varun Sharma

Acknowledgments

We would like to express our heartfelt gratitude to all the individuals and organizations who have contributed to the creation of this book, *Machining and Additive Manufacturing*, a collection of select proceedings from CPIE-2023.

First and foremost, we extend our sincere appreciation to the authors who have shared their research findings and insights in this volume. Their dedication, expertise, and scholarly contributions have enriched this book, offering valuable knowledge and advancements in the field of machining and additive manufacturing.

We would like to acknowledge the reviewers who selflessly dedicated their time and expertise to review and provide constructive feedback on the submitted papers. Their rigorous evaluation and insightful comments have played a pivotal role in ensuring the quality and academic integrity of this collection.

We are grateful to the editorial team members who worked diligently behind the scenes to manage the submission and review process, ensuring the smooth progress of the publication. Their organizational skills, attention to detail, and commitment to excellence have been instrumental in bringing this book to fruition. Nonetheless, we would also like to extend our sincere gratitude to the Springer Team for their invaluable support in the publication of this book. Their expertise and assistance have been instrumental in bringing this collection to fruition.

Our thanks also go to the conference organizers for providing a platform to foster knowledge exchange and collaboration among researchers and practitioners in the field of machining and additive manufacturing. Their efforts in organizing CPIE-2023 have facilitated the gathering of high-quality research and set the stage for the creation of this book.

More importantly, we would like to acknowledge the support and guidance of Prof. Binod Kumar Kanaujia, Director Dr. B. R. Ambedkar National Institute of Technology Jalandhar for his wholehearted support for the smooth conduct of the conference. Their commitment to promoting academic research and technological advancements is truly commendable.

Last but not least, we express our gratitude to the readers of this book. Your interest in the field of machining and additive manufacturing is what drives us to continue exploring and disseminating knowledge in this exciting area.

We hope that this book serves as a valuable resource for researchers, engineers, and professionals in machining and additive manufacturing, and inspires further advancements and collaborations in the field.

Prof. Vishal Santosh Sharma
Prof. Dr. Uday Shanker Dixit
Dr. Ajay Gupta
Dr. Rajeev Verma
Dr. Varun Sharma

Contents

About the Editors

Prof. Vishal Santosh Sharma is an honorary Associate professor at the School of Mechanical, Industrial, and Aerospace Engineering at WITS University, Johannesburg, South Africa, and a professor in the Department of Industrial & Production Engineering at Dr. B. R. Ambedkar National Institute of Technology Jalandhar, India. With over 25 years of teaching experience, he specializes in manufacturing processes/systems, industrial automation, CAD/CAM, and robotics. His research focuses on machining and additive manufacturing, including topics such as dimensional accuracy, cutting tool wear, surface roughness, and optimization using artificial intelligence algorithms. Prof. Sharma has supervised numerous Ph.D. and master's students, published over 105 papers with significant citations, and organized international conferences. He serves as an associate editor for prestigious journals and has conducted training programs with renowned experts in the field.

Prof. Uday Shanker Dixit is a distinguished Scholar and an Educator in Mechanical Engineering at IIT Guwahati. With a Ph.D. from IIT Kanpur and an M.Tech. from IIT Kanpur, he has made significant contributions to research, teaching, and leadership. He served as the head of the Department of Mechanical Engineering and the officiating director at CIT, Kokrajhar. Prof. Dixit's research interests include metal forming, manufacturing processes, and materials engineering. He has authored numerous publications and serves on editorial boards of prestigious journals. He is a respected member of professional societies such as the Indian Welding Society and has held significant positions in various academic councils and boards. Prof. Dixit's commitment to education, research, and accreditation activities has established him as a distinguished academician in the field of Mechanical Engineering.

Dr. Ajay Gupta is an Associate Professor in the Department of Industrial & Production Engineering. He holds a Ph.D. degree in Industrial Engineering and an M.Tech. in Industrial Engineering from Dr. B. R. Ambedkar NIT Jalandhar. With research interests in Operations Management, Operations Research, Theory of Constraints, Data Analytics, MIS, Materials Management, Marketing Management, and Statistics, Dr. Gupta has published more than 40 journal articles, 25 conference papers, and

5 book/chapter publications. He has organized numerous events, including conferences and short-term courses and is a Fellow of The Institution of Engineers (India). With a track record of supervising 7 Ph.D. students and guiding 23 postgraduate dissertations, Dr. Gupta has also held various administrative positions such as the head of the department, the nodal officer, the controller of examination, an associate dean, the deputy registrar, and is presently the head of Training and Placements at NIT Jalandhar.

Dr. Rajeev Verma is an Assistant Professor in the Industrial and Production Engineering Department at Dr. B. R. Ambedkar National Institute of Technology Jalandhar, Punjab, India. He obtained his Ph.D. and Masters in Engineering from the Production & Industrial Engineering Department of Punjab Engineering College, Chandigarh. With over 11 years of industrial and teaching experience, Dr. Verma's current research interests include surface engineering, thermal spray coating technology, laser surface texturing, superhydrophobic surfaces, CAE, and Taguchi methodology. He has authored and co-authored 1 Book, 10 Book Chapters, 33 papers in International Journals, 30 papers in International Conferences, and 06 papers in National Conferences. Dr Verma has guided 11 postgraduate dissertations and 11 undergraduate projects and is presently supervising 6 Ph.D. research scholars. He serves as a reviewer for prestigious scientific journals and has organized conferences, short-term courses, and internships. In addition to his accomplishments, Dr. Verma has been honored with the "Outstanding Faculty in Industrial and Production Engineering" award at the 7th Venus International Science and Technology Awards—VISTA 2021, presented by the Venus International Foundation. Furthermore, Dr. Verma has been conferred with the "National Distinguished Young Aspirant Award 2021," instituted by the International Institute of Organized Research (I2OR).

Dr. Varun Sharma is an Assistant Professor in the Department of Industrial and Production Engineering at Dr. B. R. Ambedkar National Institute of Technology Jalandhar, India (NITJ). He has previously served as an Assistant Professor in the Department of Mechanical Engineering at Shoolini University and Sant Longowal Institute of Engineering & Technology Longowal. With a background in welding engineering, Dr. Sharma's research interests span areas such as welding, materials science, physical metallurgy, tribology, additive manufacturing, and optimization through the design of experiments. He has published numerous research articles, filed a patent, and led a research project funded by TEQIP-III. Dr. Sharma has supervised several postgraduate and undergraduate projects and has been actively involved in organizing conferences and short-term courses. He is a lifetime member of the Society of Materials and Mechanical Engineers (SOMME).

Assessment of Sustainable Additive Manufacturing Drivers: A MCDM Approach

Neha Choudhary, Varun Sharma, and Pradeep Kumar

Nomenclature

AM	Additive Manufacturing
BWM	Best Worst Method
MCDM	Multi-Criteria Decision-Making

1 Introduction

The manufacturing industry is expanding very quickly and making a significant contribution to the global economy. In the current manufacturing system, increased demand, globalization, customized production, environmental concern, cost, and social responsibilities are the major issues [1]. The adoption of advanced technologies has the potential to mitigate these problems. Industry 4.0 has introduced the smart manufacturing process called Additive Manufacturing (AM). AM is a group of techniques that fabricates 3-dimensional products by adding materials in a layer-by-layer fashion in a more sustainable way [2]. Sustainable manufacturing is one of the United Nations' sustainable development goals. It is one of the measures toward sustainable development by minimizing environmental impacts while remaining economical and safe for employees and society [3].

AM is a new sustainable process being used by many industries today. It includes reduced cost, manufacturing time, tooling, assembly, resource efficiency, and provide flexibility. It also has the potential to lessen the environmental impact by producing near-net shapes. Hence, adopting sustainable AM in the manufacturing area will

N. Choudhary (✉) · V. Sharma · P. Kumar
Department of Mechanical and Industrial Engineering, Indian Institute of Technology, Roorkee 247667, India
e-mail: nchoudhary@me.iitr.ac.in

V. S. Sharma et al. (eds.), *Machining and Additive Manufacturing*, Lecture Notes in Mechanical Engineering, https://doi.org/10.1007/978-981-99-6094-1_1

provide triple-bottom-line benefits of sustainability, i.e., economic, environmental, and social [2].

Various authors have studied the benefits of AM to achieve economic, environmental, and social sustainability. The combination of life cycle assessment and life cycle costing methodology has also been studied for AM [4]. The mathematical model for the calculation of energy consumption, carbon emission, and cost involved during AM has shown its sustainable advantage [5, 6]. Few studies have also been conducted that involve the social sustainability aspect in AM [7, 8]. Additionally, various multi-criteria decision-making (MCDM) approaches have been used to select, interlink, and compare the sustainable aspects of AM. In this regard, a study has been carried out on selecting materials for AM technologies in a sustainable approach using different MCDM techniques. The selection of AM techniques considering the sustainable concept and challenges in the execution of sustainable AM has also been studied using MCDM techniques [9].

From the literature mentioned above, it can be seen that the ranking of the AM advantages in terms of sustainability has not been explored. The present work describes the setting of priority of the sustainable outcomes of AM by using an MCDM technique. The three pillars of sustainability, i.e., economic, environmental, and social, have been considered in the present work.

2 Methodology

The present work focuses on prioritizing the sustainable outcomes of AM. The proposed methodology consists of four steps, as shown in Fig. 1. First, the literature review has been performed to identify the sustainable AM drivers in economic, environmental, and social aspects as presented in Table 1. The questionnaire has been prepared and shared with the experts. The experts are selected from different fields like academics (professors and researchers), and scientists. The opinions of the experts were then gathered. The Best Worst Method (BWM) was used to determine the final ranking of the sustainable outcomes.

2.1 Best Worst Method (BWM)

The Best Worst method (BWM) is an MCDM method first proposed by Rezaei [18]. This method is selected to prioritize/rank the drivers of sustainable AM. BWM has various advantages over other MCDM techniques. It has a smaller number of pairwise comparisons, ensuring high consistency in the outcome. It involves best and worst criteria comparison, eliminating the secondary comparison matrix like other MCDM methods. The steps of BWM are as follows [18]:

1. Find the criteria (drivers) from the literature.

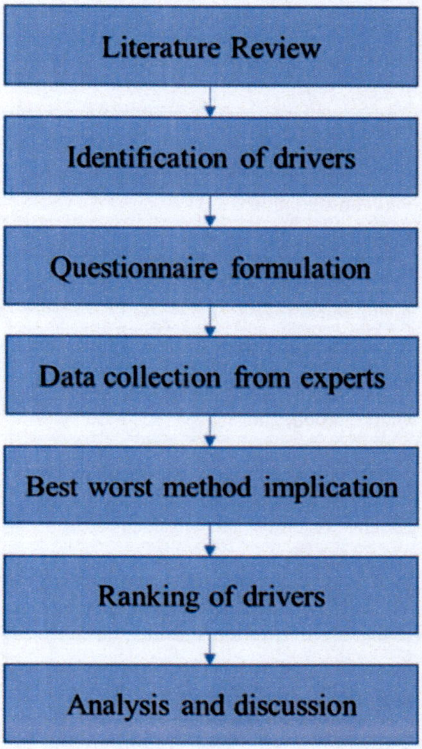

Fig. 1 Research methodology

2. Analyze the best and worst drivers with the help of experts. The best driver is the most vital driver and the worst driver is the least important driver.
3. Determine the importance of the best driver over all other drivers using a 1–9 scale. The best to others driver vector would be expressed as:

$$D_B = (d_{B1}, d_{B2}, \ldots d_{Bn}) \tag{1}$$

where d_{Bj} indicates the comparison of identified best driver B over other driver j.

4. Determine the importance of all other drivers over the worst driver using a scale of 1–9. The others to worst driver vector would be expressed as:

$$D_W = (d_{1W}, d_{2W}, \ldots d_{nW})^T \tag{2}$$

where d_{jW} implies the comparison of other drivers j over identified worst driver W.

5. Evaluate the optimal weights of each driver.

The optimal weight of each driver is one where for each pair of $\frac{W_B}{W_j}$ and $\frac{W_j}{W_w}$, we have.

Table 1 Sustainable AM drivers

Driver	References
Economic drivers	
Cost reduction (EC1)	Chandra et al. [2, 10, 11]
Just-in-time production (EC2)	Taddese et al. [11, 12]
Part consolidation (EC3)	Hohn et al. [13]
Design flexibility (EC4)	Chandra et al. [2]
Less energy and raw material use (EC5)	Chandra et al. [2]
Environmental drivers	
Less carbon and greenhouse gas emission (EN1)	El Baz et al. [10, 14]
Reduced noise production (EN2)	Taddese et al. [11]
Digital production (EN3)	Gadekar [15]
Recycling capability (EN4)	Taddese et al. [11]
Less waste production (EN5)	Chandra et al. [2]
Efficient resource consumption (EN6)	El Baz et al. [10]
Social drivers	
Customized product (S1)	Matos et al. [8, 13]
Reduced labor workload (S2)	Matos et al. [1, 7]
Reduced health risk and occupational hazard (S3)	Huang et al. [16]
Skill and carrier development (S4)	Matos et al. [8]
Increased accessibility (S5)	Hohn et al. [13, 17]

$$\frac{W_B}{W_j} = d_{Bj} \; and \; \frac{W_j}{W_w} = d_{jw}$$

To identify the weight, we have to minimize the maximum difference,

$$\min[\max_j(\left|W_B - d_{Bj}W_j\right|, \left|W_j - d_{jw}W_w\right|)] \quad (3)$$

$$s.t. \sum_j W_j = 1$$

$$W_j \geq 0, forallj$$

Equation (3) can be solved as linear optimization model,

$$\min_j \boldsymbol{\xi} \quad (4)$$

$$s.t. \left|W_B - d_{Bj}W_j\right| \leq \boldsymbol{\xi}, for\ all\ j$$

$$|W_j - d_{jw} W_w| \leq \xi, forallj$$

$$\sum_j W_j = 1$$

$$W_j \geq 0, forallj$$

On solving Eq. (4), optimum weight of the drivers can be determined. The consistency ratio of the BWM can be obtained through Eq. (5),

$$Consistency\ ratio\ (CR) = \xi / Consistency\ index\ (CI) \tag{5}$$

The consistency index can be taken from. The consistency ratio ∈ (0,1), where a value approaching toward 0 shows maximum consistency and toward 1 shows weak consistency.

For solving Eq. (4), BWM excels solver has been used.

3 Results and Discussions

The present study aims to identify and prioritize the prominent drivers of sustainable AM in economic, environmental, and social aspects. The BWM MCDM technique has been used step-by-step to analyze the selected drivers, as explained in Sect. 2.1. The three different aspects of sustainability have been taken as of similar weightage. The weight of the different drivers for each aspect has been calculated and the consistency of the result has also been evaluated. Table 2 presents the response of respondent 1 for economic drivers of AM as an example.

Table 2 Response for economic drivers (Respondent 1)

Best to other drivers	EC1	EC2	EC3	EC4	EC5
Best driver: EC5	2	4	6	3	1
Others to worst			**Worst driver: EC3**		
EC1			4		
EC2			2		
EC3			1		
EC4			4		
EC5			5		

Table 3 Weight and ranks for economic drivers of AM

Drivers	EC1	EC2	EC3	EC4	EC5
Weight	0.201444	0.100112	0.101394	**0.341565**	0.255485
Rank	3	5	4	**1**	2

3.1 Economic Drivers

The five economic drivers have been selected and ranked, as shown in Table 3. From the analysis, it was found that "Design flexibility" is the highest weighted driver, followed by "Less energy and raw material use," "Cost reduction," "Part consolidation," and "Just in time production." AM has increased the efficiency of flexible designing, unlike traditional manufacturing. In traditional manufacturing, flexibility in design is limited due to manufacturing constraints, and that too is time-consuming and expensive. AM process initiates with CAD modeling of different geometries. These geometries can be quickly modified just by changing the CAD design. This flexibility allows AM to produce difficult geometries and customized structures in less time, ultimately reducing cost and proven economic process [2]. The other economic drives are also important such as fabricating multipart assembly into a single product reduces fabrication time. The on-demand production reduces inventory and storage costs of products and raw materials. As AM is a fast process that produces near-net shape products, it reduces material and electricity consumption [11]. Hence, AM provides cost reduction that includes manufacturing and other supply chain costs.

3.2 Environmental Drivers

The ranking of environmental drivers highlights that "Less carbon and greenhouse gas emission," "Recycling capability," and "Less waste production" are the three top-ranked drivers, and "Reduced noise production" is the least important driver of AM as presented in Table 4. AM takes less time to fabricate intricate, complex-shaped products that require less electricity consumption than traditional technologies. Less electricity consumption results in fewer emissions to the environment [11]. AM can use recycled materials like in fused deposition technology of AM support structures, defective products and failed prints can be reused after shredding. It also produces less waste during manufacturing and post-processing by fabricating the product near the required dimensions, unlike traditional manufacturing techniques.

Other environmental drivers, "Digital production," "Efficient resource consumption," and "Reduced noise production," are also important drivers at their point. Digital production enables decentralized production that reduces long-distance transportation, reducing carbon footprint. AM does not require much natural resource consumption (land, water, etc.) for establishing machinery and finished products due

Table 4 Weight and ranks for environmental drivers of AM

Drivers	EN1	EN2	EN3	EN4	EN5	EN6
Weight	**0.309007**	0.072996	0.119873	0.182342	0.164525	0.151257
Rank	**1**	6	5	2	3	4

Table 5 Weight and ranks for social drivers of AM

Drivers	S1	S2	S3	S4	S5
Weight	0.291041	0.152677	**0.313999**	0.146073	0.096209
Rank	2	3	**1**	4	5

to the compact nature of machines and on-demand production. AM machinery also has less noise production advantage, unlike traditional manufacturing like forging, machining, etc. [11].

3.3 Social Drivers

The selected social drivers have covered workers' and customers' communities. It can be seen from Table 5 that "Reduced health risk and occupational hazards" has been ranked first followed by "Customized product," "Reduced labor workload," "Skill and carrier development," and "Increased accessibility." AM is an automated and green technology. It has less exposure to hazardous chemicals, dust, noise, risk, and vibrations, hence providing good working conditions to the workers. It also has the privilege of tailoring the product according to customers' needs, which results in their satisfaction. AM permits flexible and reduced working hours. It also benefits the workers to develop skills and carrier opportunities. Additionally, remote production and home fabrication enhance the accessibility of the product to the customers [17]. In this way, all other drivers are also important in social assessment.

3.4 Consistency Ratio

The Consistency Ratio (CR) measures the reliability and stability of the ranking obtained through the BWM method. The CR has been calculated using Eq. (5) for all three sustainability aspects. A low value of CR represents a high level of agreement and stability in the result. On the other hand, the high value of CR indicates less stability and reliability of the output. The value of CR exists in $0 \leq \text{CR} \geq 0.1$ range. Table 6 shows that the CR of the output lies in the defined range for all AM sustainability aspects, confirming that the ranking of the drivers is reliable.

Table 6 Overall consistency ratio for each sustainability aspect

Drivers	ξ	CI	CR
Economic	0.168002	2.3	0.073045
Environmental	0.149915	3	0.049972
Social	0.168002	5	0.028029

4 Conclusions

Sustainability has become a major concern for industries, and AM is one of the technologies that can meet that need. In this regard, the present study has used the BWM MCDM method to analyze and rank AM's selected 16 sustainability drivers with 5 economic, 5 social, and 6 environmental drivers. The weight analysis revealed the top-ranked drivers as design flexibility, less carbon and greenhouse gas emission, and reduced health risk and occupational hazards in economic, environmental, and social aspects of sustainability, respectively. The top prioritized drivers have $\geq 50\%$ agreement among the total experts. The consistency ratio confirmed the reliability of the obtained ranks, showing a value below 0.1. The analysis of AM drivers can compel industrial decision-makers to assess their willingness to deploy sustainable AM and its potential impact.

In future studies, additional drivers related to institutional and managerial aspects could be considered. The different sustainability aspects' weight could also be varied in future works. Other MCDM techniques could be used to validate the result in future.

References

1. Matos F, Godina R, Jacinto C, Carvalho H, Ribeiro I, Peças P (2019) Additive manufacturing: exploring the social changes and impacts. Sustain 11:3757
2. Chandra M, Shahab F, KEK V, Rajak S (2022) Selection for additive manufacturing using hybrid MCDM technique considering sustainable concepts. Rapid Prototyp J 28:1297–1311
3. Agrawal R, S. V (2020) Sustainability evaluation of additive manufacturing processes using grey-based approach. Grey Syst Theory Appl 10:393–412
4. Ingarao G, Priarone PC (2020) A comparative assessment of energy demand and life cycle costs for additive-and subtractive-based manufacturing approaches. J Manuf Process 56:1219–1229
5. Peng T, Kellens K, Tang R, Chen C, Chen G (2018) Sustainability of additive manufacturing: an overview on its energy demand and environmental impact. Addit Manuf 21:694–704
6. Zavadskas EK, Govindan K, Antucheviciene J, Turskis Z (2016) Hybrid multiple criteria decision-making methods: a review of applications for sustainability issues. Econ Res Istraz 29:857–887
7. Naghshineh B, Ribeiro A, Jacinto C, Carvalho H (2021) Social impacts of additive manufacturing: a stakeholder-driven framework. Technol Forecast Soc Change 164:120368
8. Matos F, Jacinto C (2019) Additive manufacturing technology: mapping social impacts. J Manuf Technol Manag 30:70–97
9. Alsaadi N (2021) Prioritization of challenges for the effectuation of sustainable additive manufacturing: a case study approach. Processes 9

10. El Baz J, Tiwari S, Akenroye T, Cherrafi A, Derrouiche R (2022) A framework of sustainability drivers and externalities for industry 4.0 technologies using the best-worst method. J Clean Prod 344:130909
11. Taddese G, Durieux S, Duc E (2020) Sustainability performance indicators for additive manufacturing: a literature review based on product life cycle studies. Int J Adv Manuf Technol 107:3109–3134
12. Ford S, Despeisse M (2016) Additive manufacturing and sustainability: an exploratory study of the advantages and challenges. J Clean Prod 137:1573–1587
13. Hohn MM, Durach CF (2021) Additive manufacturing in the apparel supply chain—impact on supply chain governance and social sustainability. Int J Oper Prod Manag 41:1035–1059
14. Top N, Sahin I, Mangla SK, Sezer MD, Kazancoglu Y (2022) Towards sustainable production for transition to additive manufacturing: a case study in the manufacturing industry. Int J Prod Res, 1–22
15. Gadekar R, Sarkar B, Gadekar A (2022) Key performance indicator based dynamic decision-making framework for sustainable industry 4.0 implementation risks evaluation: reference to the Indian manufacturing industries. Ann Oper Res 318:189–249
16. Huang SH, Liu P, Mokasdar A, Hou L (2013) Additive manufacturing and its societal impact: a literature review. Int J Adv Manuf Technol 67:1191–1203
17. Rayna T, Striukova L (2016) From rapid prototyping to home fabrication: how 3D printing is changing business model innovation. Technol Forecast Soc Change 102:214–224
18. Rezaei J (2015) Best-worst multi-criteria decision-making method. Omega 53:49–57

Exploring Sustainable Manufacturing: A Comprehensive Review of Literature and Practices

SantanKumar Chaurasiya and Gurraj Singh

1 Introduction

Manufacturing has been a crucial component of global economic growth for decades, but this progress has come at a significant cost to the environment. The traditional manufacturing practices have led to increased greenhouse gas emissions, depletion of natural resources, and waste generation. The need for a more sustainable approach to manufacturing has become more apparent in recent years, with climate change and environmental degradation posing a significant threat to our planet's future [1]. Sustainable manufacturing is an emerging field that focuses on reducing the environmental impact of manufacturing processes while maintaining economic growth. The concept of sustainable manufacturing involves integrating sustainability considerations into all aspects of the manufacturing process, from the sourcing of raw materials to the disposal of end products. The goal is to minimize the use of non-renewable resources, reduce waste generation, and promote environmentally friendly production practices [2, 3]. The concept of sustainable manufacturing has gained significant attention from various stakeholders, including governments, companies, and consumers. Sustainable manufacturing practices have become a critical business strategy for companies looking to reduce their environmental footprint, meet regulatory requirements, and improve their brand image. The trend toward sustainable manufacturing is also driven by consumer demand for environmentally friendly products and the desire to support socially responsible companies [4].

The principles of sustainable manufacturing are not only limited to environmental considerations but also encompass social and economic factors. The concept of sustainable manufacturing involves a balance between economic growth, social responsibility, and environmental sustainability. Sustainable manufacturing practices

S. Chaurasiya (✉) · G. Singh
Department of Industrial and Production Engineering, Dr B R Ambedkar National Institute of Technology, Jalandhar 144011, India
e-mail: chaurasiyasantan@gmail.com

V. S. Sharma et al. (eds.), *Machining and Additive Manufacturing*, Lecture Notes in Mechanical Engineering, https://doi.org/10.1007/978-981-99-6094-1_2

must ensure that workers' rights are protected, and the local communities benefit from the manufacturing process [5]. The implementation of sustainable manufacturing practices involves a significant shift in the manufacturing paradigm, which requires the integration of sustainability considerations into all levels of decision-making. Sustainable manufacturing involves not only the adoption of new technologies and processes but also requires the development of new business models and supply chain management practices [6]. The importance of sustainable manufacturing is reflected in the United Nations' Sustainable Development Goals (SDGs). SDG 9, Industry, Innovation, and Infrastructure, specifically highlight the need for sustainable manufacturing practices. The goal is to promote inclusive and sustainable industrialization and foster innovation, while reducing the environmental impact of manufacturing processes [7].

The transition toward sustainable manufacturing requires a collaborative effort from all stakeholders, including companies, governments, non-governmental organizations (NGOs), and consumers. Companies can lead the way by implementing sustainable manufacturing practices that not only reduce their environmental footprint but also create value for their customers and stakeholders. Governments can play a significant role in promoting sustainable manufacturing practices by implementing policies and regulations that incentivize companies to adopt sustainable practices. NGOs and consumer groups can also raise awareness of sustainable manufacturing practices and encourage companies to adopt environmentally and socially responsible practices [8–10].

2 Sustainable Manufacturing Practice

Sustainable manufacturing practice refers to the implementation of manufacturing processes that minimize negative impacts on the environment, while also promoting economic development and social well-being. This involves reducing waste and pollution, conserving natural resources, adopting renewable energy sources, and using sustainable materials. Sustainable manufacturing also involves incorporating social and ethical considerations into the design and production process, such as fair labor practices, human rights, and community engagement. The article "Sustainable Manufacturing: Metrics, Standards, and Infrastructure" by Rachuri et al. [11] aim to provide an overview of the current state of sustainable manufacturing, including its metrics, standards, and infrastructure [11]. The article highlights the importance of measuring sustainability in manufacturing and provides an overview of the most commonly used metrics, including life cycle assessment (LCA) and carbon footprint. The authors also discuss the various standards that are used to evaluate sustainability, including ISO 14001 and the Global Reporting Initiative (GRI). The article further delves into the infrastructure necessary for sustainable manufacturing, including technology and workforce development [12]. The authors highlight the importance of technological advancements in improving the sustainability of manufacturing processes, such as the use of renewable energy and the development of closed-loop manufacturing systems. They also discuss the need for a well-trained workforce with the necessary skills to implement sustainable manufacturing practices [13]. The article "Sustainable manufacturing: A bibliometric analysis" presents a bibliometric analysis of sustainable manufacturing research over a 30-year period from 1990 to 2019. The study aims to identify the intellectual structure, influential publications, and emerging trends in sustainable manufacturing research. The results show that sustainable manufacturing research has grown significantly in recent years, with a noticeable increase in publications since 2010. The intellectual structure of the field is diverse, with a wide range of topics and themes being studied, such as circular economy, ecodesign, and green supply chain management. The study identifies several influential publications and authors, as well as emerging research trends such as Industry 4.0 and sustainable production systems [14].

Machado et al. [15] focus on the role of innovation in promoting sustainable manufacturing practices. They highlight the importance of innovation in driving sustainable manufacturing practices and provide insights into the potential opportunities and challenges associated with promoting sustainability through innovation. The study is also adding valuable contribution to the field of sustainable manufacturing, and provides useful insights for policymakers, industry practitioners, and researchers [15]. The article by Miah & Ryan (2018) presents a study on the relationship between innovation and sustainability in manufacturing companies. The findings suggest that innovation positively impacts sustainability in manufacturing companies. They found that companies that adopted environmentally friendly technologies and practices were more likely to achieve sustainability goals. Additionally, they found that companies that emphasized social sustainability were more likely to achieve economic sustainability. Finally, they found that companies that invested in sustainable practices were more likely to be innovative.

The article Habidin et al. [16] provides a comprehensive review of sustainable manufacturing practices in the automotive industry, including energy efficiency, material efficiency, waste reduction, and supply chain sustainability. The authors argue that the automotive industry is a significant contributor to environmental degradation and social issues, and that sustainable manufacturing practices are critical for addressing these challenges [16]. The article by Kumar and Mani [17] presents a review of sustainability assessment methodologies used in manufacturing. The authors review a range of assessment methodologies used in the manufacturing sector, including life cycle assessment (LCA), sustainability performance measurement (SPM), sustainable value stream mapping (SVSM), and sustainability index (SI). They highlight the strengths and limitations of each methodology and discuss how they can be used in the manufacturing sector to support sustainable development. They also suggest that future research could focus on developing integrated assessment methodologies that consider the interactions between environmental, social, and economic factors [17]. The article by Bendig et al. [18] aims to provide a systematic review of the role of government policies and regulations in promoting sustainable manufacturing practices. The article starts with a brief introduction to sustainable manufacturing practices and the importance of government policies in promoting them. The authors reviewed 53 research articles and identified several key government policies and regulations that have been successful in promoting sustainable manufacturing practices. These include ecolabeling and environmental standards, economic instruments, and regulatory frameworks. The article also discusses the challenges that governments face in implementing these policies and regulations, such as a lack of political will and resistance from industry stakeholders [18]. The article by Wasserbaur et al. [19] aims to provide a systematic review of the role of government policies and regulations in promoting sustainable manufacturing practices [19]. The article starts with a brief introduction to sustainable manufacturing practices and the importance of government policies in promoting them. The authors reviewed 53 research articles and identified several key government policies and regulations that have been successful in promoting sustainable manufacturing practices. These include ecolabeling and environmental standards, economic instruments, and regulatory frameworks. The article also discusses the challenges that governments face in implementing these policies and regulations, such as lack of political will and resistance from industry stakeholders [18].

The article "Circular Economy in Manufacturing: A Review" explores the concept of circular economy and its potential application in manufacturing. The authors highlight that the traditional linear model of manufacturing, which involves extracting resources, making products, using them, and then disposing of them, is no longer sustainable due to limited resources and environmental concerns. They argue that a circular economy, which emphasizes reducing waste and reusing resources, can provide a more sustainable approach to manufacturing [20].

3 Sustainable Supply Chain Management (SSCM) and Manufacturing Practice in Business Performance

The article reviews the literature on sustainable manufacturing practices (SMPs) and their impact on business performance. The authors conducted a systematic literature review to identify and analyze the existing literature on SMPs and business performance. They identified 84 studies published between 2000 and 2018 and analyzed them using a thematic approach. The review finds that SMPs have a positive impact on business performance, including economic, environmental, and social performance. The economic benefits include cost savings, increased revenue, and improved competitiveness. The environmental benefits include reduced energy consumption, waste reduction, and decreased emissions. The social benefits include improved health and safety, employee satisfaction, and community engagement [21]. The article "Building Information Modelling for Sustainable Construction" by Gao et al. provides an overview of Building Information Modelling (BIM) technology and its role in sustainable construction practices. The authors highlight the potential of BIM technology to improve the sustainability of the construction industry by facilitating more efficient use of resources, reducing waste and emissions, and enhancing collaboration and communication among project stakeholders [22].

The article titled "Enhancing sustainability in supply chains through inter-firm collaborations: An empirical investigation" aims to explore the relationship between inter-firm collaboration and sustainability in supply chains. The study was conducted through a survey of 169 Indian manufacturing firms, and the data was analyzed using Structural Equation Modelling (SEM). The study highlights the importance of collaboration in achieving sustainability goals and emphasizes the need for policymakers to encourage such collaboration through appropriate policies and incentives. Overall, the study contributes to the literature on sustainable supply chains by providing insights into the role of inter-firm collaboration in enhancing sustainability [23].

In this article, Bovea and Pérez-Belis review the literature on models for sustainable supply chain management (SSCM). They provide an overview of various approaches to SSCM, such as life cycle assessment, ecodesign, and closed-loop supply chains, and discuss their strengths and weaknesses. The authors also examine the role of stakeholders in SSCM and the importance of collaboration and information sharing among supply chain partners. The article concludes with a discussion of future research directions in SSCM, including the need for more empirical research, the development of integrated SSCM models, and the use of emerging technologies such as blockchain and the Internet of Things. The authors also emphasize the importance of considering social sustainability issues in addition to environmental concerns, and the need for SSCM models to be adapted to the specific context and goals of each supply chain [24]. In another study provided by Panigrahi et al. [25] a comprehensive review of the existing literature on sustainable supply chain management (SSCM) and identifies areas for future research. The article highlights the need for a more integrated and collaborative approach to SSCM, involving all stakeholders in the supply chain. The authors suggest that future research should focus on developing frameworks and tools to enable organizations to implement SSCM effectively. They also call for more empirical studies to evaluate the impact of SSCM on organizational performance and sustainability outcomes [25].

4 Sustainable Manufacturing Practice: Challenges and Future Trend

The article "Sustainable manufacturing education and training: A review of current practices and future trends" presented by Domadi MK [26], authors conclude that sustainable manufacturing education and training should be integrated into existing engineering and business curricula to ensure that graduates have the necessary knowledge and skills to address sustainability challenges. They also recommend the development of new programs and the use of innovative teaching methods, such as experiential learning and industry collaborations. Additionally, the authors emphasize the importance of lifelong learning and professional development to ensure that professionals stay up to date with the latest sustainable manufacturing practices [26]. The article "Sustainable Production: A Critical Review" by Colicchia et al. examines the state of research on sustainable production, identifying gaps and opportunities for future research. They suggest that future research should aim to develop new models and frameworks that can better capture the complex interactions between these different factors. Additionally, the authors highlight the importance of engaging stakeholders and creating new business models that enable sustainable production practices to be integrated into the broader value chain [27]. In this article, the authors provide a comprehensive review of the literature on sustainable manufacturing systems (SMS). The study examines research trends and identifies gaps in the field, providing a roadmap for future research. The authors highlight the importance of sustainable manufacturing systems in achieving environmental sustainability, economic development, and social well-being. The article first presents the evolution of sustainable manufacturing and then discusses the key elements of SMS such as design for sustainability, green supply chain management, lean manufacturing, and life cycle assessment. The article also discusses various drivers and barriers of SMS implementation and explores the role of government policies and regulations in promoting sustainable manufacturing [28]. The article by Ocampo and Clark [29] presents a framework for developing a sustainable manufacturing and operations strategy. The framework provides a comprehensive perspective that considers environmental, social, and economic factors, as well as stakeholder engagement and collaboration, The article concludes by highlighting the need for further research to address the challenges associated with sustainable manufacturing modeling and optimization. The authors argue that there is a need for more integrated approaches that can capture the interrelationships between product, process, and system levels, and that can account for the complex trade-offs between different sustainability objectives [29].

The article by Jayal et al. [30] presents a comprehensive review of the literature on sustainable manufacturing, and highlights some of the key research trends and directions in the field. The authors argue that there is a need for more integrated approaches that can capture the interrelationships between product, process, and system levels, and that can account for the complex trade-offs between different sustainability objectives [30]. The article "Integrating sustainability into operations management research and practice: Recent progress and future directions" by Wu, Cegielski, and Hazen aims to examine the recent progress in integrating sustainability into operations management (OM) research and practice and provide future research directions. The study identified that sustainability is becoming increasingly integrated into OM research and practice, with a growing number of studies investigating the linkages between sustainability and OM. The review also identified gaps in current research, such as a lack of focus on the impact of sustainable practices on OM performance and a lack of attention given to the trade-offs between sustainability and other OM objectives. The authors suggest future research should focus on measuring the impact of sustainable practices on OM performance and identify trade-offs between sustainability and other OM objectives [31]. The article "Achieving sustainability through small business initiatives: Exploring the role of information technology" by Luederitz et al. [32] reviews the literature on how small businesses can achieve sustainability through the adoption of information technology (IT) initiatives. The authors identified six IT initiatives that can promote sustainability in small businesses: cloud computing, social media, big data analytics, mobile devices, radio frequency identification (RFID), and green IT. The authors then discuss the mechanisms through which these initiatives can promote sustainability, including reducing energy consumption, promoting waste reduction, and improving supply chain management [32].

We discussed a comprehensive review of the literature on sustainable manufacturing and operations management, identifying gaps and opportunities for future research. The authors highlight the importance of integrating sustainability into existing curricula, developing new programs, and using innovative teaching methods. They also emphasize the need for more integrated approaches that can capture the interrelationships between product, process, and system levels, and account for the complex trade-offs between different sustainability objectives. Additionally, the articles discuss the role of government policies, stakeholder engagement, and the adoption of information technology initiatives in promoting sustainable practices.

5 Conclusion

Sustainable manufacturing is a critical component of the global effort to promote environmental sustainability, social responsibility, and economic growth. The field of sustainable manufacturing has gained significant attention from various stakeholders, including governments, companies, and consumers. Sustainable manufacturing practices involve integrating sustainability considerations into all aspects of the manufacturing process, from the sourcing of raw materials to the disposal of end products.

The implementation of sustainable manufacturing practices requires a significant shift in the manufacturing paradigm, which requires the integration of sustainability considerations into all levels of decision-making. The transition toward sustainable manufacturing requires a collaborative effort from all stakeholders, including companies, governments, NGOs, and consumers. The importance of sustainable manufacturing is reflected in the United Nations' Sustainable Development Goals, which specifically highlight the need for sustainable manufacturing practices.

Overall, sustainable manufacturing practices have become a critical business strategy for companies looking to reduce their environmental footprint, meet regulatory requirements, and improve their brand image. Sustainable manufacturing practices not only reduce the environmental impact of manufacturing processes but also create value for customers and stakeholders. The transition toward sustainable manufacturing requires a collective effort from all stakeholders, and continued research and innovation in this area can have a significant impact on environmental sustainability and social responsibility.

6 Future Scope

The future scope of sustainable manufacturing is promising, with many opportunities for improvement and growth. Companies that prioritize sustainable manufacturing practices will be better positioned to meet the evolving demands of consumers and regulators and contribute to a more sustainable future.

References

1. Rosen MA, Kishawy HA (2012) Sustainable manufacturing and design: concepts, practices and needs
2. Opferkuch K, Caeiro S, Salomone R, Ramos TB (2021) Circular economy in corporate sustainability reporting: a review of organisational approaches. Bus Strateg Environ 30:4015–4036. https://doi.org/10.1002/bse.2854
3. Chaurasiya S, Singh G (2023) Life cycle assessment of nanocomposite manufactured using ultrasonic stir casting. J Mater Sci 58:5298–5318. https://doi.org/10.1007/s10853-023-08363-0
4. Gupta DS, Dangayach G, Posinasetti N (2018) Implementation of sustainable manufacturing practices in Indian manufacturing companies. Benchmarking An Int J 25:2441–2459. https://doi.org/10.1108/BIJ-12-2016-0186
5. Bastas A (2021) Sustainable manufacturing technologies: a systematic review of latest trends and themes
6. Felsberger A, Reiner G (2020) Sustainable industry 4.0 in production and operations management: a systematic literature review
7. Jansen LJM, Kalas PP (2020) Improving governance of tenure in policy and practice: a conceptual basis to analyze multi-stakeholder partnerships for multi-stakeholder transformative governance illustrated with an example from South Africa
8. Abubakr M, Abbas AT, Tomaz I, Soliman MS, Luqman M, Hegab H (2020) Sustainable and smart manufacturing: an integrated approach. Sustain 12:19. https://doi.org/10.3390/su12062280
9. Fofou RF, Jiang Z, Wang Y (2021) A review on the lifecycle strategies enhancing remanufacturing
10. Garza-Reyes JA, Villarreal B, Kumar V, Ruiz P (2016) Lean and green in the transport and logistics sector–a case study of simultaneous deployment. Prod Plan Control 27. https://doi.org/10.1080/09537287.2016.1197436
11. Rachuri S, Sriram R, Narayanan A, Sarkar P, Lee J, Lyons K, Kemmerer S (2010) Sustainable manufacturing: metrics, standards, and infrastructure-workshop summar
12. Chaurasiya S, Singh G (2023) Sustainability assessment comparison of cutting fluid for turning of titanium alloy grade II. Process Integr Optim Sustain. https://doi.org/10.1007/s41660-023-00322-1
13. Sudarsan R, Sriram RD, Narayanan A, Sarkar P, Lee JH, Lyons KW, Kemmerer SJ (2010) Sustainable manufacturing: metrics, standards, and infrastructure-workshop summary. In: 2010 IEEE international conference on automation science and engineering. pp 144–149
14. Bhatt Y, Ghuman K, Dhir A (2020) Sustainable manufacturing. Bibliometrics and content analysis. J Clean Prod 260:120988. https://doi.org/10.1016/j.jclepro.2020.120988

15. Machado CG, Winroth MP, Hans E, Ribeiro D, Gonçalves C, Winroth MP, Hans E, Ribeiro D (2020) Sustainable manufacturing in industry 4. 0 : an emerging research agenda. Int J Prod Res 7543:24. https://doi.org/10.1080/00207543.2019.1652777
16. Habidin NF, Zubir A, Fuzi N, Md Latip NA, Azman MNA (2015) Sustainable manufacturing practices in Malaysian automotive industry: confirmatory factor analysis. J Glob Entrep Res 5:1–13. https://doi.org/10.1186/s40497-015-0033-8
17. Kumar M, Mani M (2022) Sustainability assessment in manufacturing for effectiveness: challenges and opportunities. Front Sustain 3. https://doi.org/10.3389/frsus.2022.837016
18. Bendig D, Kleine-Stegemann L, Gisa K (2023) The green manufacturing framework—a systematic literature review. Clean Eng Technol 13:100613. https://doi.org/10.1016/j.clet.2023.100613
19. Wasserbaur R, Sakao T, Milios L (2022) Interactions of governmental policies and business models for a circular economy: a systematic literature review. J Clean Prod 337:130329. https://doi.org/10.1016/j.jclepro.2021.130329
20. Bjørnbet MM, Skaar C, Fet AM, Schulte KØ (2021) Circular economy in manufacturing companies: a review of case study literature. J Clean Prod 294:126268. https://doi.org/10.1016/j.jclepro.2021.126268
21. Abdul-Rashid SH, N. S., Raja Ghazilla RA, Ramayah T (2017) The impact of sustainable manufacturing practices on sustainability performance: empirical evidence from Malaysia. Int J Oper Prod Manag 37:182–204. https://doi.org/10.1108/IJOPM-04-2015-0223
22. Haruna A, Shafiq N, Montasir OA (2021) Building information modelling application for developing sustainable building (multi criteria decision making approach). Ain Shams Eng J 12:293–302. https://doi.org/10.1016/j.asej.2020.06.006
23. Alzoubi H, Ahmed G, Al-Gasaymeh A, Al Kurdi B (2019) Empirical study on sustainable supply chain strategies and its impact on competitive priorities: the mediating role of supply chain collaboration. Manag Sci Lett 10. https://doi.org/10.5267/j.msl.2019.9.008
24. Seuring S (2013) A review of modeling approaches for sustainable supply chain management. Decis Support Syst 54:1513–1520. https://doi.org/10.1016/j.dss.2012.05.053
25. Panigrahi S, Bahinipati B, Jain V (2018) Sustainable supply chain management: a review of literature and implications for future research. Manag Environ Qual Int J 30. https://doi.org/10.1108/MEQ-01-2018-0003
26. Domadi MK (2016) A review on current trend of sustainable manufacturing in Malaysia
27. Geels FW, McMeekin A, Mylan J, Southerton D (2015) A critical appraisal of sustainable consumption and production research: the reformist, revolutionary and reconfiguration positions. Glob Environ Chang 34:1–12. https://doi.org/10.1016/j.gloenvcha.2015.04.013
28. Chan F, Li N, Chung S-H, Saadat M (2016) Management of sustainable manufacturing systems-a review on mathematical problems. Int J Prod Res 55:1–16. https://doi.org/10.1080/00207543.2016.1229067
29. Ocampo L, Clark E (2015) A sustainable manufacturing strategy framework: the convergence of two fields. 20:29–57
30. Jayal AD, Badurdeen F, Dillon OW, Jawahir IS (2010) Sustainable manufacturing: modeling and optimization challenges at the product, process and system levels. CIRP J Manuf Sci Technol 2:144–152. https://doi.org/10.1016/j.cirpj.2010.03.006
31. Walker H, Klassen R, Sarkis J, Seuring S (2014) Sustainable operations management: recent trends and future directions. Int J Oper Prod Manag 34. https://doi.org/10.1108/IJOPM-12-2013-0557
32. Luederitz C, Caniglia G, Colbert B, Burch S (2021) How do small businesses pursue sustainability? the role of collective agency for integrating planned and emergent strategy making. Bus Strateg Environ 30:3376–3393. https://doi.org/10.1002/bse.2808

Performance Analysis of 3D Printed Impellers for Portable Vacuum Cleaner

Check for updates

Om Prakash Mishra, Surender Singh, Krishan Kumar, and Rakesh Kumar

1 Introduction

3D printer is a new technology for the fabrication of prototypes and various models, which are difficult to manufacture from traditional methods. This not only increases the ease of getting the prototype model fabricated but also creates the production process simple [1]. This paper illustrates the performance measurement of a portable vacuum cleaner (PVC) using three different impellers. This PVC will be used to clean PCB, keypad, and other electronic circuitry, small household purposes, computer motherboard, DVD player, sofa, bookshelf, desktop, where accessibility is a challenge. To carry out the performance check, designing and development of PVC has been carried out in an additive manufacturing lab. All the parts of PVC have been printed using a 3D printer of FDM technology. The three different types of impellers, i.e., open, semi-closed, and closed impellers have also been printed. This vacuum cleaner uses an impeller to suck the dust particles while cleaning operation. The assembling and test running have been carried out to check the operation ability. The performance of PVC has been modeled by changing the impeller. The smart features in the vacuum cleaner have been introduced such as, easy to operate, effective for small purposes, compact design, and rechargeable (so, no wiring connection needed). The various types of impellers, outer body designs, casing, slots, holders, seats for various components, blade fans, etc., have been developed with the help of a 3D printer. 3D printed parts stand quite robust, if designed well and give equal performance as any manufactured component [2]. The cleaning ability of any vacuum cleaner (VC) is not just about vacuum suction power, but other parameters are equally important such as; power consumption, noise level, vibration, etc. Dust particles are sucked by air, flowing from the opening at the cleaning head or tool, through the

O. P. Mishra (✉) · S. Singh · K. Kumar · R. Kumar
Department of Mechanical Engineering, J C Bose University of Science and Technology, YMCA, Faridabad 1210006, India
e-mail: opmishra@jcboseust.ac.in

V. S. Sharma et al. (eds.), *Machining and Additive Manufacturing*, Lecture Notes in Mechanical Engineering, https://doi.org/10.1007/978-981-99-6094-1_3

main body, the bag and/or filter system, and then out the exhaust port. This airflow is created by the vacuum motor, which also may be referred to as the suction motor. The motor consists of various electrical components attached to a fan or multiple fans called impellers. When the fans spin, a partial vacuum is created and the pressure inside the vacuum cleaner drops below the ambient (or existing) air pressure in surrounding cleaning area. Because air pressure is higher outside the vacuum cleaner than inside, air rushes through the vacuum cleaner [3]. The paper is organized into six sections; Introduction, literature review, methodology, description of 3D printed PVC, setting up experiment, performance analysis, and conclusion.

2 Literature Review

3D printed parts can be effectively used to design and analysis of any equipment after then further courses of correction are needed to make it commercialize [4]. By using a 3D printer to print all parts of vacuum cleaner except its motor and have demonstrated it working well. This research paper also focuses on assembles of different components and the proper working of fused filament fabrication 3D printers. Structural analysis of mini drones developed using 3D printing techniques can be commercialized [4]. Modeling of complicated design of the impeller of a Centrifugal Compressor on 3D-printing machine using FDM technology is advantageous for several arrangements of the machining process [5, 6]. 3D printing for industrial applications is growing at a fast rate, which is saving a lot of cost, energy, effort, and human involvement [7]. Therefore, it is found that 3D printed components are easy to fabricate, alter designs, and giving similar results in industrial applications.

2.1 Problem Formulation

Vacuum cleaner works effectively with good performance of impellers. Impeller is a device used to increase the pressure of fluid and is greatly influenced by the design of it. There are three types of impellers available which are open, semi-closed, and closed impellers. Out of the three important impellers, which one will give the best performance if it is fitted in a portable vacuum cleaner?

3 Methodology

The methodology of this representation (Fig. 1) is as follows.

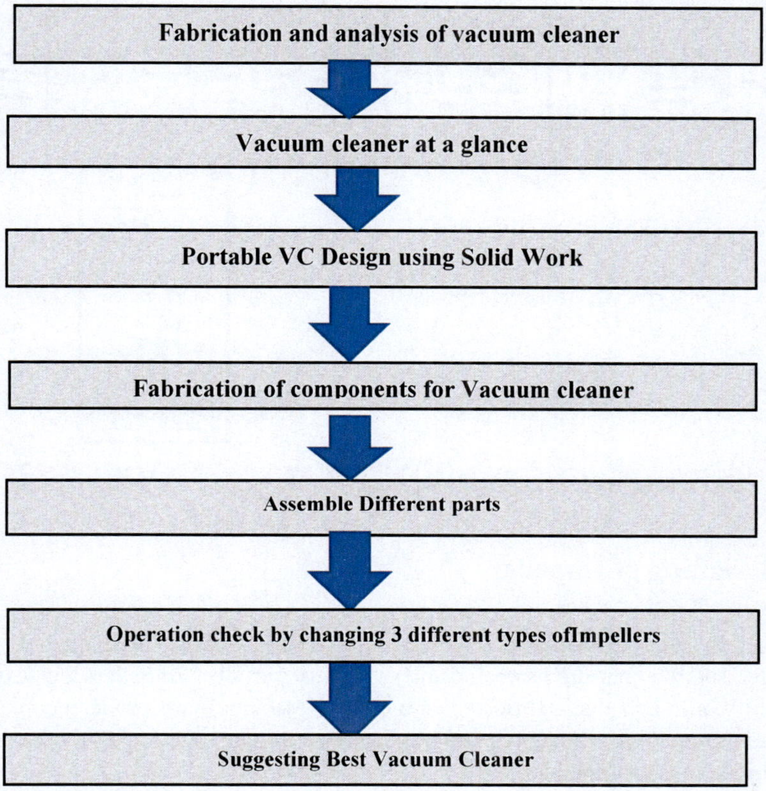

Fig. 1 Methodology to carry out performance analysis

1. The method adopted here is to design, fabricate, assemble, and test the vacuum cleaner. The various parts of the vacuum cleaner will be printed using an FDM 3D printer. The performance of PVC will be checked experimentally by fitting the three impellers alternately.
2. Comparative study of PVC performance by using analytic hierarchical process (AHP) will be done to verify the result obtained experimentally.

4 3D Printed VC at a Glance

The schematic diagram of VC is shown in Fig. 2. The other printed parts are not shown in this paper.

The main components of PVC are cover, dust collector, impeller, motor, exhaust fan, filtering mechanism, control panel, and battery. The fitment of each component is shown in Fig. 2. As this paper is focusing on design variations of impellers, therefore, a brief discussion on impeller design is only mentioned here.

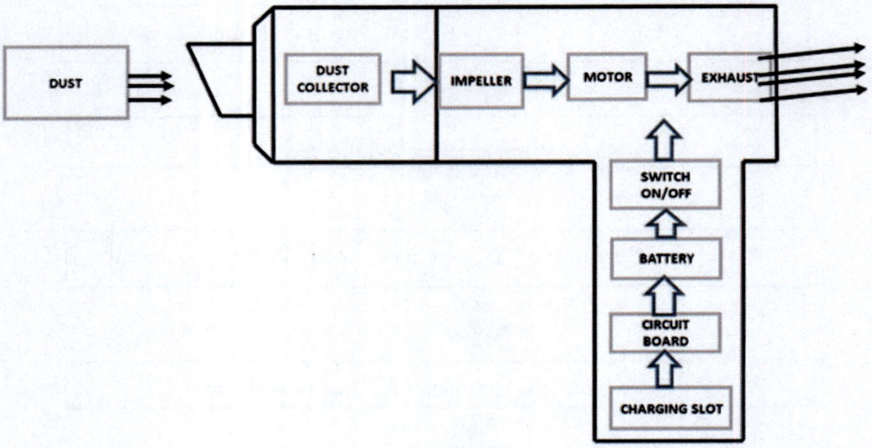

Fig. 2 Schematic diagram of vacuum cleaner

5 Designing of Impeller

Impeller plays an important role in the vacuum cleaner because, with the help of an impeller, suction pressure is created and sucks dust particles. Impellers are designed in solid works and then fabricated in an FDM machine with proper geometrical accuracy. Initially, the design parameters were set to fabricate impellers. We chose here three types of impellers.

- **Open Impeller**

Open impeller creates low suction pressure as compared to closed impeller and semi-closed impeller. Many dust particles are not sucked by an open impeller. The strength of an open impeller is weak. An open impeller is shown in Fig. 3.

- **Semi-closed impeller**: Practically, semi-closed impeller creates more suction pressure as compared to an open impeller and it creates less suction pressure as compared to a closed impeller, but it does not clean dust particles properly. Blade of semi-closed impeller is supported on one side of the wall as shown in Fig. 4.
- **Closed impeller**: Practically, a closed impeller sucks light dust particles and it creates more suction pressure as compared to semi-closed impeller and open impeller. Blades of closed impeller is supported back and front side walls as shown in Fig. 5.

Fig. 3 Open impeller

Fig. 4 Semi-closed impeller

6 Setting up Experiment

After assembling of PVC, a simple experiment was performed to check the varying performance of this by replacing the three impellers. We carried out noting down the important parameters of a PVC performance based upon the fitted impellers out of the three, i.e., open, semi-closed, and closed impellers. The following three

Fig. 5 Closed impeller

parameters were noted for each of the impeller-fitted VC. Three parameters were checked with each impeller, i.e., noise level, sucking capacity, and battery running hours of PVC. The noting down of the below parameter has been performed in the additive manufacturing lab, where a clean environment was available. The value of three observations in each category is shown in Tables 1, 2, and 3. For checking these parameters, a PC keyboard was selected. Small pieces of paper and sand were placed on the board and VC was operated to such these items. We noted down noise level in dB with each impeller. Similarly, sucking capacity was checked in gm/s and then the running time of VC till the battery is completely exhausted. The battery is chargeable and after each charge, experiment was performed with the other two impellers fitted.

Table 1 Noise level

S. No	Impeller	Value in decibel
1	Closed	62
2	Semi-closed	63
3	Open	65

Table 2 Sucking capacity

S. No	Impeller	Value in gm/sec
1	Closed	0.1515
2	Semi closed	0.0224
3	Open	0.0143

Table 3 Running time

S. No	Impeller	Running time in min. for a 1200mAH battery
1	Closed	43
2	Semi-closed	41
3	Open	40

6.1 Experimental Analysis

Noise Level: Vacuum cleaners generally operate at a noise level **between 56 and 80 dB**. A full-sized vacuum cleaner outputs a noise level of 70–80 dB, which is extremely loud and is the same as a garbage disposal. We find the best noise level is of closed impeller, i.e., 62dB, which is quite below the loudness limit. Table 1 shows the result.

Sucking Capacity

A "Vacuum Test Gauge" has been used for measuring the suction power. This actually measures the displacement of water, which says more about the suction power of a vacuum cleaner. The gauge measured the best result from VC fitted with a closed impeller. It can suck up to 0.15g/s. Table 2 shows the result.

Running Time

To measure the power consumption of a vacuum cleaner with different impellers, a 1200mAH battery is used to calculate the running time for single battery usage. This feature has been checked with the capacity of the current drawing from the battery. We find closed impeller consuming minimum wattage and therefore, its running time is the highest. Table 3 shows the result.

6.2 AHP Analysis

The above parameters show the best result given by a closed impeller. But the overall result will be valid, when vacuum cleaner is also checked on some general categories such as the material of the casing of PVC, dust collecting performance, operational noise level, aesthetic look, and power consumption. AHP has been used extensively in different areas of decision-making, even in complex situations as an effective tool of MCDM [8]. It has emerged as a very effective tool because the pair-wise comparison makes the procedure impervious to comparison inaccuracies. The methodology is quite suitable and simple to meet the objective of this study. Here, comparison of three PVCs with fitted impellers is to be carried out on general perceptions as mentioned above. Thus, comparison may also solidify our experimental result obtained. The AHP model to find an optimum performing VC is shown in Fig. 6.

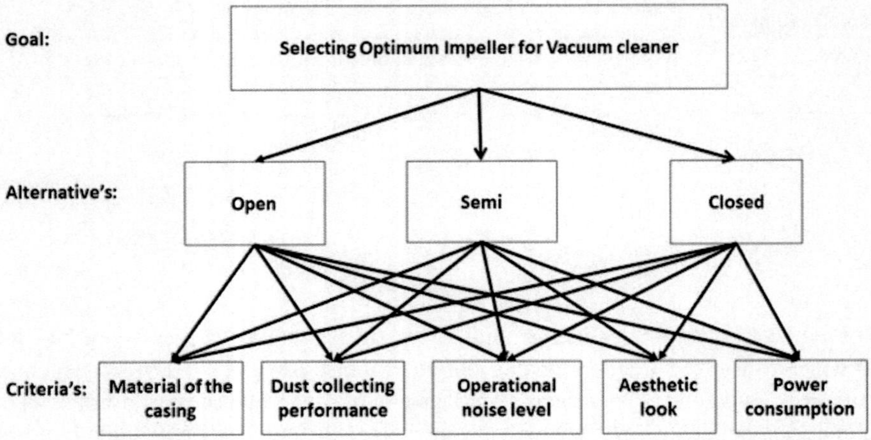

Fig. 6 AHP model to find optimum performing VC

Overall Priority Vector

	MC	DCP	ONL	AL	PC	Overall priority vector
OPEN	0.127	0.284	0.273	0.238	0.103	0.2317
SEMI	0.512	0.096	0.086	0.621	0.664	0.2925
CLOSED	0.36	0.619	0.638	0.136	0.23	**0.4685**

Based upon the above observation, it is found that PVC fitted with closed impeller is best performing if all other variables are fixed. Regarding its sustainability, the author clearly stated that 3D printed material can well sustain vibration and dynamic forces [1].

7 Conclusion

This paper illustrated the performance analysis of a PVC based on the performance of the three-impeller designed in-house in an additive manufacturing lab. The complete parts of the PVC have been printed using 3D printer of FDM technology. The filament used is ABS material. The impeller being a very important component of any VC, this paper undertook to measure its performance using three different shaped impellers, i.e., open, semi-closed, and closed. These three impellers were fitted one by one to measure three parameters i.e.; noise level, sucking particle diameter, and battery running hours. The experiment was performed and observation was noted. Measurement of these parameters shows that closed impeller is the best performing in all the three categories. To verify this result, we used AHP considering general criteria such as; dust collecting performance, operational noise level, aesthetic look, and power consumption. The AHP also figures out the closed impeller as best performing. This paper may be very useful for academicians and industrialists to understand the importance of the shape of impeller fitted in any vacuum cleaner as well as several parameters, which are necessary to examine for various usages.

Limitation: The paper is written based upon in-house experiment by measuring only three parameters of performance of tis VC. Whereas more parameters such as flow rate, dynamics of flow, etc., can also be measured using CFD (Computation Fluid dynamics).

References

1. Joshi SC, Sheikh AA (2015) 3D printing in aerospace and its long-term sustainability. Virtual Phys Prototyp 10(4):175–185
2. Mirón V, Ferràndiz S, Juàrez D, Mengual A (2017) Manufacturing and characterization of 3D printer filament using tailoring materials. Procedia Manuf 13:888–894
3. Nirupan R, Pothiraj S, Roopan P, Uthranarayan C (2019) Floor cleaning robot with 3D printed vacuum impeller and casing. Int Res J Eng Technol 6(2):1082–1086
4. Krishna, Gundeti A (2017) A comparative study on the components fabricated by injection moulding and fdm 3d printing process. Int J Mech Prod Eng 5(12):116–121
5. Muralidharan N, Pratheep VG, Shanmugam A, Hariram A, Dinesh P, Visnu B (2021) Structural analysis of mini drone developed using 3D printing technique. Mater Today: Proc 46:8748–8752
6. Sandeep DC, Chhabra D (2017) Comparison and analysis of different 3d printing techniques. Int J Latest Trends Eng Technol 8:264–272
7. Jandyal A, Chaturvedi I, Wazir I, Raina A, Haq MIU (2022) 3D printing–a review of processes, materials and applications in industry 4.0. Sustain Oper Comput 3:33–42
8. Saaty TL (1980) The analytic hierarchy process, McGraw-Hill, New York.

3D Printed Electronics: Role of Materials and Processes

P. Akashraj, A. Maria Joseph Mervin, Harjot Chane, R. K. Lishwanth, Md Ahsanul Habib Rahat, Panjul Garg, Vishal Francis, and Narendra Kumar

1 Introduction

As technology grows in rapid phases, the process of developing electronic components is also expanding. 3D printing has made it easier to fabricate electronics and also plays a major role in innovation. The global electronics components market is expected to rise continuously by 7.7% as a result of the current electronics boom, from 5.41 million US dollars in 2017 to 7.86 million US dollars in 2022 [1]. The most common method for producing printed electronics is to use printing processes frequently used in the graphics sector to impart functional inks to plates (screen, ink-jet, flexo, gravure, etc.) [2]. There are several methods of 3D printing electronic components or circuits. They may be contact type or noncontact type. Majorly, the 3D printing techniques are noncontact in nature. The materials used are widely pursued since they offer numerous advantages in terms of ease of processing, good compatibility with a variety of substrates, and a great opportunity for structural modification [3]. In recent years, various printing technologies have been developed in the areas of screen printing, inkjet printing, and microcontact printing. Most of the work has been for the development of an inkjet solution capable of processing conductive material. The biggest potential benefit for high-density, high-speed, miniature innovative packaging is offered by nanomaterials, composites, and hybrids. These structures are incredibly unique materials with a wide range of potential uses due to their small dimensions, strength, and remarkable physical and electrical properties. With the utilization of 3D printing for electronic components, the application area is enlarged

P. Akashraj · A. Maria Joseph Mervin · H. Chane · R. K. Lishwanth · M. A. H. Rahat · P. Garg · V. Francis (✉)
School of Mechanical Engineering, Lovely Professional University, Jalandhar, India
e-mail: vishal.24813@lpu.co.in

N. Kumar
Department of Industrial and Production Engineering, Dr. B R Ambedkar National Institute of Technology, Jalandhar 144027, India

V. S. Sharma et al. (eds.), *Machining and Additive Manufacturing*, Lecture Notes in Mechanical Engineering, https://doi.org/10.1007/978-981-99-6094-1_4

[4]. There have been efforts to develop post-processing procedures for 3D printed parts to give them the appropriate electronic properties to get around the high cost and constraints of specific 3D printed electronic systems employing surface direct write technologies. The most well-known of these is probably electroplating plastic components, which is done after a preliminary metallization by electroless deposition or conductive paint [2]. The application area of 3D printed electronics covers stretchable electronics, radio frequency antennae, and 3D structural electronics [5–23]. The present paper discusses the recent advances in 3D printing techniques for printed electronics. Moreover, the materials required for electronics applications are also discussed along with the applications areas of 3D printed technology.

2 Techniques for Printed Electronics

2.1 Conventional Techniques for Printed Electronics

There is direct contact between the print head and the intended substrate when using conventional electronic fabrication techniques known as contact printing to transfer the required inked patterns. The functional ink is normally transferred to the target substrate by contact printing procedures using a single sheet or engraved roll. However, the benefits and drawbacks of various contact printing methods could differ based on the printing principles used, therefore critical to carefully compare how each contact printing method operates. There are two steps in the printing process for gravure printing. During the first stage, an ink fountain's inks are collected by a gravure roller, and the amount of ink that is applied is controlled by a blade. Similarly, an ink-blocking stencil prevents an unwanted part of the substrate from touching the functional ink.

Rotary Screen printing uses pressure from an impression roller to imprint predetermined patterns on the substrate. Flexographic printing (also known as offset printing), in contrast to gravure printing and rotary screen printing, uses a soft plate roller to transport a controlled measure of ink. The pattern is indirectly imprinted on the substrate and the screen-printing technique variation known as flatbed screen printing recently allured a lot of attention.

2.2 Screen Printing

Screen printing can be applied in many ways, including Roll-to-Roll (R2R) production with an optimum solution and printing settings. Using an optimal approach and the same process constraints, screen printing can be done again and again with identical results [24].

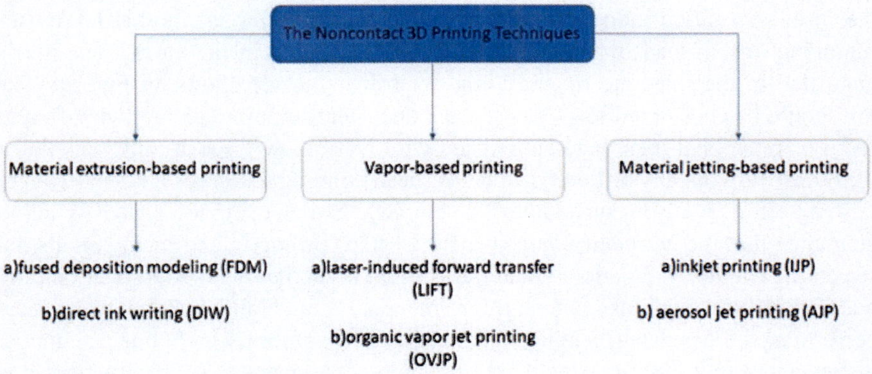

Fig. 1 Classification of noncontact type printing

For screen printing, typically, flatbed screen printing involves coating the screen with a printable solution, coating the mesh on the stencil with ink using a squeegee, and then pressure applied in a controlled manner. High-speed production is made possible by rotary system screens, but they are costly and need frequent cleaning owing to material clogging. However, several factors, including the geometry of the squeegee angle, the mesh size, the solution viscosity, and the snap-off between the screen and the substrate, have an impact on the quality of screen printing.

2.3 3D Printing Techniques for Printed Electronics

Regardless of the various benefits, the printing quality and setup costs of contact printing limit its application for the fabrication of electronics. Noncontact 3D printing helps to create adaptable and affordable electronic device applications. The noncontact 3D printing method, in contrast, printing may generate less waste and reach a greater resolution. As a result, it has received considerable interest in several areas.

The noncontact 3D printing techniques are further categorized into three types as illustrated in Fig. 1.

2.4 Material Extrusion-Based Printing

In this material extrusion-based printing, the filament is fabricated using the hot extruder head and the filament is made of thermoplastic material shown in Fig. 3a. In the working principle of the fused deposition modeling (FDM) technique, the melted filament is extruded through a nozzle. Though fused deposition modeling is used in printed electronics, the application of the printed electronics is very confined because of resolution and printing quality. Direct ink writing (DIW) is similar to the fused deposition modeling (FDM) technique, the difference is that DIW uses functional ink instead of filament, and the ink is stored in the barrel. The ink is deposited in the substrate by extruding out from a nozzle shown in Fig. 3b. The ink should be considered low-viscosity ink and shear-thinning ink are two different types of functional inks that can be used with DIW. Low-viscosity ink runs down the nozzle without any difficulty, whereas shear-thinning ink needs a much stronger pushing effort. A curing system must be employed when using low-viscosity ink to guarantee that the extruded filament material fixes promptly and maintains its 3D structure. The result functional material is often combined with additional binding material in functional inks, which might reduce the functional material's electrical performance. Dispensing techniques come in a variety of forms, including pneumatic, piston, and screw. To guarantee, the printed filament creates a steady structure; it is essential to apply the right printing parameters and conditions.

2.5 Vapor-Based Printing

The modified technique of laser direct-writing (LDW) known as laser-induced forward transfer (LIFT) has been used to deposit materials to use for electronic applications. This method concentrated laser pulse with high intensity as employed to deliver energy enough to heat the thin film interface. High-quality electrical designs can be created by vaporizing metal from the melted solid film interface and depositing it on a target substrate because of energy accumulation. Organic vapor jet printing (OVJP) method utilizes heated cells that house the molecular organic source material to create saturated source vapor that is carried by a hot inert carrier gas. After expanding, the gas mixture impinges on the substrate creating a high-resolution line.

2.6 Material Jetting-Based Printing

There is growing engrossment in adopting material jetting-based printing for manufacturing electronics, because of the benefits of higher precision and less waste. Inkjet printing (IJP) can be divided into two categories based on different operating principles: continuous inkjet (CIJ) printing systems and drop-on-demand (DoD) printing systems. The printed pattern is produced by the uncharged drops while the charged ones are reused. In systems based on DoD, the droplets are ejected. An innovative noncontact ink writing method called AJP is used to create electronic circuits instead of IJP. Using a carrier gas flow, the functional ink is first atomized into droplets (2–5 m) and then transferred to the print head. AJP has the potential to print with highly viscous inks and even on curved surfaces maintain a good resolution.

2.7 *Inkjet Printing*

Using a nozzle to deposit drops which may be coupled to a basin of liquid material, inkjet printing is a type of digital printing. The resolution level is determined by the droplet's volume and area covered as it leaves the nozzle. Additionally, it is reliant on how the substrate wets. When it comes to reducing floor space, the initial investment, and the commissioning time, inkjet or digital printing facilities are a preferred option among other printing equipment. A production line can be scaled up to produce electronic prototypes using simultaneous printing on a large number of printers.

Due to their superior conductivity and endurance (several thousands of operating hours), inorganic/metal inks are most frequently utilized for inkjet printing. The affordable materials can be utilized for single-use applications like radio frequency identification (RFID) antennas, which are often employed as distinguishing tags for inventory control in retail outlets or to expedite payment processes. Continuous inkjet (CIJ) and drop-on-demand (DOD) printing techniques are the two primary inkjet printing subcategories (Figs. 2 and 3).

2.8 *Aerosol Jet Printing*

Aerosol jet printing (AJP) provides a rapid way to create miniature features on the substrate material. The technique does not require a mask, therefore, sometimes referred to as maskless material deposition. Various electronic components and devices can be printed with this technique. Moreover, many materials can be used in this process such as materials required for the fabrication of dielectric and conductive traces. In contrast to conventional electronics production or commonly used direct-write technologies, a key component of AJP facilitates the development of a wide range of devices with higher geometric complexity.

The technique can be used for fabricating intricate features and even solar cells and screens. A wide range of material viscosities can be used in AJP and printing can be done on even nonplanar surfaces. The gas flow that results from atomization or ultrasonication produces a spray of tiny droplets that power the aerosol printing process. Because it can handle, AJP is regarded as the most intriguing technology among other contact-less printing methods since it can print on various materials with greater resolution designs. Figure 3 illustrates the comparison of printing techniques.

1) Continuous Inkjet	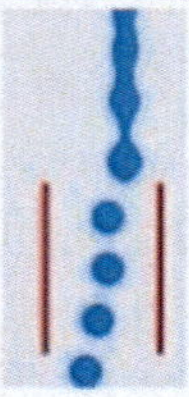	A stream of ink is continuously ejected by a nozzle, and charged droplets are deflected and positioned using an electric field. Remaining Droplets are collected by gutter and recycled.
2)Drop on Demand Inkjet		Droplets are ejected on demand. It is a simpler system compared to continuous inkjet
2a	Thermal	The ink is heated by a heating element and ejected as droplets through the nozzle.
2b	Piezo	A piezoelectric transducer is applied with a voltage pulse which creates an acoustic wave propagation in the ink and droplets are ejected through the nozzle.
2c	Electrostatic	The droplet formation involves a complex interaction of surface tension ratio between ink and nozzle and electric field. The signal fed to the printhead balances forces to create ink droplets.

Fig. 2 Categorization of inkjet printing

3 Material for 3D Printed Electronics

Printing conductive materials has gained the most interest among these methods for the manufacture of complex 3D electronics due to the durability, affordability, and development of new printers. Printing enables the construction of intricate electrical circuits in a 3D way, in contrast to conventional microfabrication techniques that often produce flat devices. When using different 3D printing methods for electronics manufacture, knowledge of the spectrum of materials and accendibility, or from prototype to manufacturing, are crucial factors to take into account. Understanding the characteristics of the existing materials for electronic manufacturing and the technologies employed is essential to comprehending the future function of 3D printing processes. These fundamental electronic parts are constructed from functional materials including metal, semiconductors, dielectrics, and polymers that may be connected, printed, or physically positioned on a variety of substrates using

Main types	Fabrication process	Fabrication techniques	Material types
• Gravure printing	• Contact printing	• Roll-to-Roll printing	• Ink-based
• Flexographic printing	• Contact printing	• Roll-to-Roll printing	• Ink-based
• Offset printing	• Contact printing	• Roll-to-Roll printing	• Ink-based
• Rotary screen printing	• Contact printing	• Roll-to-Roll printing	• Ink-based
• Flatbed screen printing	• Contact printing	• Sheet based printing	• Ink-based
• Fused deposition modeling	• Noncontact printing	• 3D printing	• Filament-based
• Direct ink writing	• Noncontact printing	• 3D printing	• Filament-based
• Organic vapor jet printing	• Noncontact printing	• Vapor deposition	• Vapor-based
• Laser-induced forward transfer	• Noncontact printing	• Laser direct write	• Vapor-based
• Aerosol jet printing	• Noncontact ink writing	• 3D printing	• Ink-based
• Inkjet printing	• Noncontact ink writing	• 3D printing	• Ink-based
• Hybrid printing	• Noncontact printing	• 3D printing	• Multi-material based

Fig. 3 Comparison of printing techniques for printed electronics [25]

various printing and patterning methods. The materials for 3D printed electronics can be solids, inks, pastes, etc.

3D printed electronics can be made using various materials such as:

1. Conductive Filaments: filaments that contain conductive materials such as metals or carbon-based composites.
2. Resins: photopolymer resins that contain conductive materials or nanoparticles.
3. Metal Powders: metal powders such as copper or nickel that can be sintered or fused using a laser or other heat source to create conductive paths within a 3D printed structure.
4. Graphene and Carbon Nanotubes: These materials can be added to resins or filaments to enhance their conductive properties.
5. Silicone Rubber: silicone rubber for flexible and stretchable electronics.

The choice of material depends on the specific requirements of the 3D printed electronics such as flexibility, conductivity, and resistance to heat or chemicals. To be processed in the same layer as other functional, dielectric, and photo-imageable materials, printable materials must be physically stable toward them. Devices and packages must maintain their structural and electrical integrity and be sturdy to provide a longer service life. Since organic and polymeric materials have several benefits, such as processing under low temperatures, durability, and a large possibility for change in structure, organic materials have been extensively sought for these uses. Nanomaterials are also a great choice for 3D printed electronics because of the biggest potential advantage for high speed, miniature innovative bundling that is offered by nanomaterials, composites, and hybrids. These structures are extremely unusual materials with a wide range of intriguing uses because of their tiny dimensions, strength, and exceptional physical and electrical characteristics. Functional polymer, composite, and hybrid-based semiconductor devices are regarded as good choices for applications involving electronics, where they completely use printable polymer technologies and various fillers may be added to a functional polymer system. Advanced packaging uses have been described for several nanocomposites. There is room for improvement in the current materials, allowing for the development of low processing temperature, adaptable, and affordable printable processes, as well as materials for mass production, even though some of the nanocomposites used to advance semiconductor packaging technology are not always printable.

3.1 Conductive Polymer Filaments

Over the past two decades, conductive inks have attracted a growing amount of interest and are transforming the industry. Mainly because of their qualities including conductivity, appropriateness for printing substrates, ease of processing, and mechanical flexibility, as well as for their capacity to impart new traits, prowess, and intricate functions.

For use in electronics, a wide range of materials, including organic and inorganic materials as well as conductors and semiconductors, have been investigated. Water, oil, or solvent-based inks are the most popular varieties. The ink's basic composition consists of various substances both in solid and liquid-based materials, each with unique properties that are tailored to the characteristics of printing technology, allowing the ink to be applied to a wide range of applications with ease. Because conductive nanoscale particles are included in the mix, conductive inks are electrically conductive. Metallic nanoparticles are frequently stabilized in ink solutions by organic ligand shells or capping agents. This creates a uniform and stable dispersion and prevents particles from sticking to one another [25].

3.2 Semiconductors

Semiconductors adjust their electrical characteristics in reaction to environmental factors such as electric fields, mechanical strains, and chemical adsorptions. However, under electrical control, they can also change their physical properties. Semiconductors are necessary for many different electronic devices, including transistors and sensors, because of these characteristics. Even though silicon- or oxide-based inorganic semiconductor materials have traditionally been used in high-performance electronics by the photolithography-based subtractive fabrication process, organic semiconductor materials have been actively researched as 3D printable inks due to their easy deformability and solubility to various organic solvents [26].

In general, inorganic semiconductors perform better and are more stable than organic semiconductors. As a result, research was done on creating printable ink using inorganic semiconductor materials. Dynamic 3D structures are made possible by semiconducting inks' manipulation of viscosity.

3.3 *Dielectric*

Electrically insulating materials have been made easier to use as dielectric materials for transistors and capacitors, or as substrates on which other electronics may be built. Electronics' structural variety and integrity have recently been shown to increase with the use of 3D printing for insulating materials. Electrically insulating materials have been made easier to use as dielectric materials for transistors and capacitors, or as substrates. Electronics' structural variety and integrity have recently been shown to increase with the use of 3D printing for insulating materials. Despite having a lower dielectric constant than inorganic materials, polymers have been intensively explored as 3D printed inks because of their low processing temperature properties. Particularly, the brittleness of inorganic materials can become a restriction in flexible/ soft electronics, while organic materials can get around this problem. The use of hybrid ink made of organic–inorganic substances has recently been described in investigations for 3Dprintable dielectric layers.

3.4 *Metallic Nanoparticle*

They have unique physical, chemical, and biological properties that make them promising materials for 3D printed electronics. For example, nanoparticles can be used as conductive inks in 3D printing to create electronic devices with high conductivity and improved mechanical properties. Additionally, nanoparticles can be used in the development of 3D printed sensors, actuators, and energy storage devices. Nanoparticle-based 3D printing has the potential to revolutionize the way electronics are manufactured and enable the creation of new and innovative devices. The individual metallic nanoparticles in metallic nanoparticle inks are enclosed in a layer of insulating organic additives and stabilizing agents and suspended in liquid media. In addition to preventing the metallic nanoparticles from agglomerating, the organic additives and stabilizing agents also obstruct the passage of electrons from the particles to the surface [27]. Figure 4 shows some of the applications of nanocomposite in 3D printed electronics.

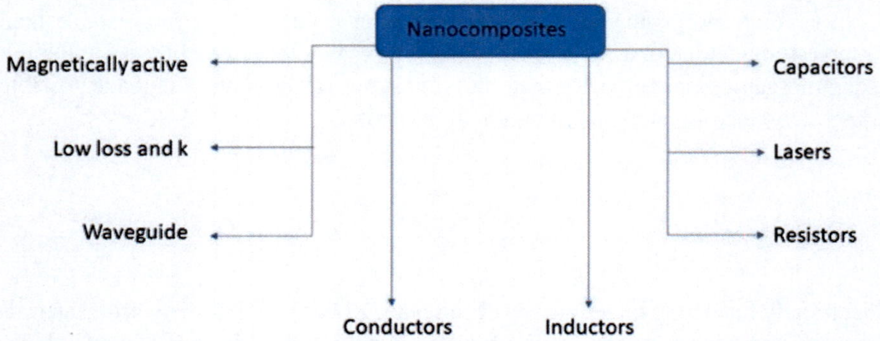

Fig. 4 Overview of some of the potential applications of nanocomposite in 3D printed electronics [27]

3.5 *Carbon Nanotubes and Graphene*

Because of their unique traits, carbon nanotubes (CNTs) are still one of the most researched materials in use today. They find use in wearable electronics, energy storage and harvesting devices, optoelectronics, chemical sensors, and water filters. Conductive inks and an accumulation process make use of several printing methods, including screen printing, transfer printing, and contact printing, creating conductive patterns on flexible substrates like carbon nanotubes. The conductive inks used in printing techniques are mainly based on volatile solutions containing conducting or semiconducting micro- and nanoscale materials, such as carbon nanotubes (CNTs), graphene, conducting polymer nanoparticles, metallic nanorods, or a mix of this material [3]. Because of exceptional mechanical qualities, particularly electrical conductivity, CNT-based inks in particular have become an attractive contender for the construction of malleable electronics utilizing printing technology. Additionally, the adaptability of CNTs has made it possible to print a variety of devices, including flexible actuators, supercapacitors, sensors, and transistors.

The ink is suitable for screen printing, dip coating, and traditional printing methods. Additionally, it has two-step 3D printing capabilities, which would make it simple to fabricate patterned conductive structures on a variety of flexible substrates without subjecting them to difficult processing conditions.

4 Applications of 3D Printed Electronics

The innovative technology of 3D printing commonly referred to as additive manufacturing or AM, has been around for quite some time but has recently gained more and more popularity. 3D printed electronics offer great potential for building complex objects with multiple features.

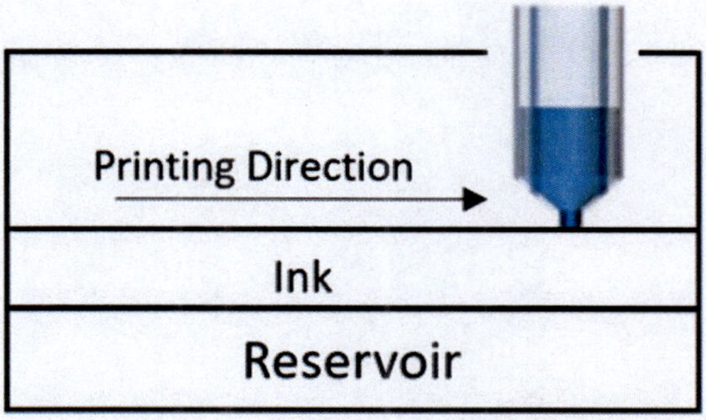

Fig. 5 The e-3DP procedure is illustrated schematically [30]

3D printed electronics have been presented as the next additive frontier. In previous years, a large amount of research and effort related to 3D printed electronics has been carried out by universities and industry. Due to its layer-by-layer or point-by-point properties, 3D printing encourages the integrated assembly and embedding of additional components. Sensors, electronics, and embedded components, among others, are incorporated into 3D printed goods. In printed electronics, metal-based particle-containing solutions or suspensions are used to print contacts and interconnects. Metal nanoparticles, often known as nanoparticle inks, are employed in ink formulations in the suspension form [28, 29].

4.1 *Stretchable Electronics*

Stretchable electronics are a brand-new class of electronic devices made possible by recent interest in wearable electronics and soft robotics (Fig. 5). e-3DP's invention of incredibly flexible sensors opens up new options for manufacturing soft-functioning gadgets for wearable electronics, soft robotics, and other applications. It makes it possible for electrical devices to have conformance, a lightweight design, and shock-resistant construction, which are difficult to achieve when utilizing rigid substrates like glass plates and semiconductor wafers [30].

4.2 *Radio Frequency Antenna*

AM has potential use in the construction of RF antennas (Fig. 6). Due to material and processing limitations, conventional antennas are rarely conformal or flexible, but AM makes it possible to create antennas that are more effective, lighter in weight, and smaller in size. Because UAVs are frequently used for high-risk missions and certain UAVs are not designed to be retrieved at all, the advantages of AM in terms of faster fabrication times and lower costs become increasingly alluring for UAV

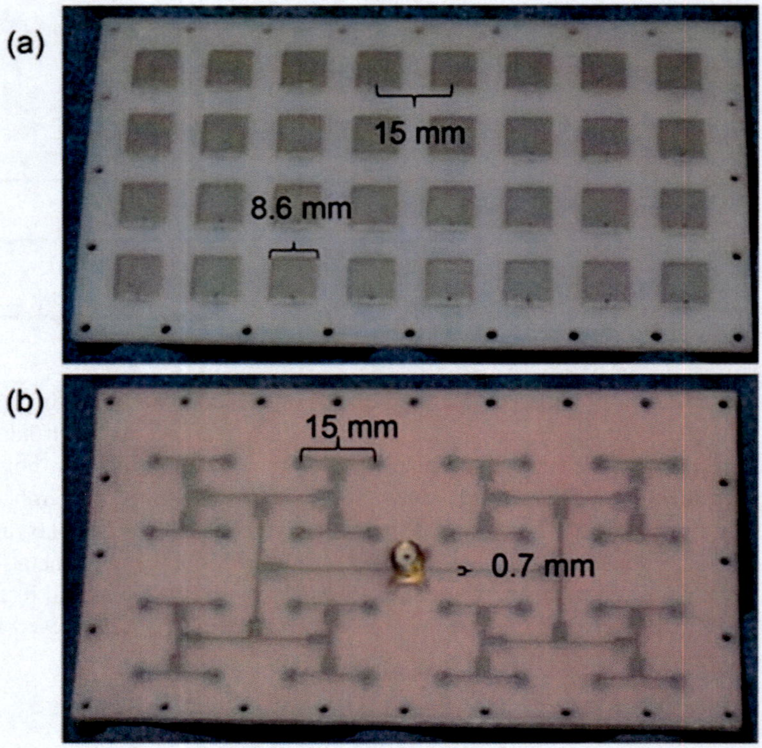

Fig. 6 Antenna radiating patch parts made additively and **b** feed network with SMA connector [31]

antennas. Increased installation sites, reduced weight, smaller size, and lower cost are made possible by this impact of aerodynamics on the airborne platform.

Figure 6 shows a conformal AM antenna as an illustration. An existing UAV platform's skin was used in the design of this antenna to ensure a smooth fit. Using a poly jet printer, the substrate material was 3D-produced and had a variety of curvatures throughout the sample. The micro-dispense nozzle is kept evenly spaced from the substrate throughout the printing process thanks to the printer's utilization of the coordinates established by laser mapping. This guarantees that the final print will have the greatest resolution available. This technique was used to print the feed network, the vias linking the two, and the radiating patch elements [31].

4.3 3D Structural Electronics

A PμSL (projection micro stereolithography) method is one method used to construct 3D structures. Combining hybrid manufacturing and additive manufacturing (AM)

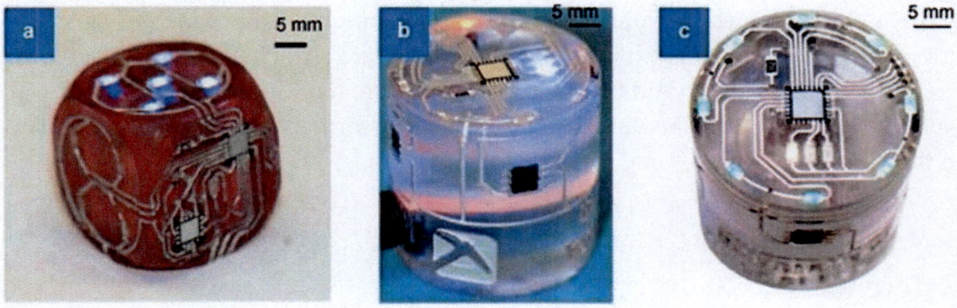

Fig. 7 **a** A game die with a microcontroller and an accelerometer. **b**, **c** A magnetometer system that includes a microcontroller and orthogonal Hall Effect sensors [25]

technologies will enable fully 3D, high-resolution, multi-material, and large-area fabrication. The conductive tracks were created by combining DWC (direct-write/cure) with CNT/polymer nanocomposites, which could pave the way for a new generation of affordable 3D structural electronics in the field of consumer, defense, and medical electronics (Fig. 7). The use of various viscous materials in a single build might cause contamination problems, and there are still difficulties with throughput, limited materials, and conductive inks that can cure at low temperatures [25].

5 Conclusions

3D printed electronics offer great potential for the fabrication of electronic components and devices. Compared to traditional techniques 3D printed techniques offer numerous advantages for printed electronics. Utilizing the benefits of 3D printing techniques has opened new avenues in the field of printed electronics. Moreover, the development of viable materials for the fabrication of electronic components and devices has further paved the way for the utilization of the 3D printing approach for printed electronics. One of the main benefits of 3D printing electronics is the ability to create highly customized and optimized devices. This customization can lead to better performance and efficiency compared to off-the-self-help components. Additionally, not all materials are suitable for electronic components, so careful selection is important. Despite these challenges, 3D printing electronics has the potential to revolutionize the way we design and manufacture electronic devices, especially for small-scale production and prototyping. As the technology continues to improve and mature, we will likely see even more innovative applications of 3D printing in electronics. It can be concluded that 3D printed electronics have revolutionized the way we think about manufacturing and production. It offers numerous benefits over traditional methods, such as speed, customization, and cost-effectiveness. The ability to print complex electronic components and circuits with precision and accuracy has opened new possibilities in various fields, including healthcare, aerospace, and consumer electronics. Additionally, the technology has the potential to reduce waste and environmental impact by allowing for on-demand production and minimizing the need for excess inventory. While there are still challenges to overcome, such as improving the speed and quality of printing, the future looks bright for 3D electronic printing, and it is sure to continue to transform the manufacturing industry in the years to come.

References

1. Espera AH, Dizon JRC, Chen Q, Advincula RC (2019) 3D-printing and advanced manufacturing for electronics. Prog Addit Manuf 4(3):245–267
2. Flowers PF, Reyes C, Ye S, Kim MJ, Wiley BJ (2017) 3D printing electronic components and circuits with conductive thermoplastic filament. Addit Manuf 18:156–163
3. Das RN, Lin HT, Lauffer JM, Markovich VR (2011) Printable electronics: towards materials development and device fabrication. Circuit World 37(1):38–45
4. Maalderink HHH, Bruning FBJ, de Schipper MMR, van der Werff JJJ, Germs WWC, Remmers JJC, Meinders ER (2018) 3D printed structural electronics: embedding and connecting electronic components into freeform electronic devices. Plast, Rubber Compos 47(1):35–41
5. Wiklund J, Karakoç A, Palko T, Yiğitler H, Ruttik K, Jäntti R, Paltakari J (2021) A review on printed electronics: fabrication methods, inks, substrates, applications, and environmental impacts. J Manuf Mater Process 5(3):89
6. Rigon D, Ricotta M, Meneghetti G (2020) A literature survey on the structural integrity of 3D printed virgin and recycled ABS and PP compounds. Procedia Struct Integr 28:1655–1663
7. Yan C, Yang X, Wu H (2023) 4D printing of stimuli-responsive materials. Addit Manuf, 85–112
8. Saengchairat N, Tran T, Chua C-K (2017) A review: additive manufacturing for active electronic components. Virtual Phys Prototyp 12(1):31–46
9. Lau G-K, Shrestha M (2017) Ink-jet printing of micro-electro-mechanical systems (MEMS). Micromachines 8(6):194
10. Smith M, Choi YS, Boughey C, Kar-Narayan S (2017) Controlling and assessing the quality of aerosol jet printed features for large area and flexible electronics. Flex Print Electron 2(1):015004
11. Lin C-T, Hsu C-H, Chen I-R, Lee C-H, Wu W-J (2011) Enhancement of carrier mobility in all-inkjet-printed organic thin-film transistors using a blend of poly(3-hexylthiophene) and carbon nanoparticles. Thin Solid Films 519(22):8008–8012
12. de Gans B-J, Duineveld PC, Schubert US (2004) Inkjet printing of polymers: state of the art and future developments. Adv Mater 16(3):203–213
13. Li M, Li Y-T, Li D-W, Long Y-T (2003) Recent developments and applications of screen printed electrodes in environmental assays—a review. Anal Chim Acta 734:31–44
14. Cruz SMF, Rocha LA, Viana JC (2018) Printing technologies on flexible substrates for printed electronics. Flex Electron
15. Elschner A, Kirchmeyer S, Lovenich W, Merker U, Reuter K (2010) PEDOT: principles and applications of an intrinsically conductive polymer. CRC Press
16. Kim K, Kim G, Lee BR, Ji S, Kim S-Y, An BW, Song MH, Park J-U (2015) High-resolution electrohydrodynamic jet printing of small-molecule organic light-emitting diodes. Nanoscale 7(32):13410–13415
17. Kimura Y, Nagase T, Kobayashi T, Hamaguchi A, Ikeda Y, Shiro T, Takimiya K, Naito H (2015) Soluble organic semiconductor precursor with specific phase separation for high performance printed organic transistors. Adv Mater 27(4):727–732
18. Kwon J, Takeda Y, Shiwaku R, Tokito S, Cho K, Jung S (2019) Three-dimensional monolithic integration in flexible printed organic transistors. Nat Commun 10(1):54

19. Ortiz RP, Facchetti A, Marks TJ (2010) High- *k* organic, inorganic, and hybrid dielectrics for low-voltage organic field-effect transistors. Chem Rev 110(1):205–239
20. Jung S, Kim JH, Kim J, Choi S, Lee J, Park I, Hyeon T, Kim D-H (2014) ReverseMicelle-induced porous pressure-sensitive rubber for wearable human-machine interfaces. Adv Mater 26(28):4825–4830
21. Shin SR, Farzad R, Tamayol A, Manoharan V, Mostafalu P, Zhang YS, Akbari M, Jung SM, Kim D, Comotto M, Annabi N, Al-Hazmi FE, Dokmeci MR, Khademhosseini A (2016) A bioactive carbon nanotube-based ink for printing 2D and 3D flexible electronics. Adv Mater 28(17):3280–3289
22. Lu B-H, Lan H-B, Liu H-Z (2018) Additive manufacturing frontier: 3D printing electronics. Opto-Electron Adv 1(1):17000401–17000410
23. Molesa S, Redinger DR, Huang DC, Subramanian V (2003) High-quality inkjet-printed multi-level interconnects and inductive components on plastic for ultra-low-cost RFID applications. MRS Proc 769:1–6
24. Palavesam N, Marin S, Hemmetzberger D, Landesberger C, Bock K, Kutter C (2018) Rollto-roll processing of film substrates for hybrid integrated flexible electronics. Flex Print Electron 3(1):014002
25. Tan HW, An J, Chua CK, Tran T (2019) Metallic nanoparticle inks for 3D printing of electronics. Adv Electron Mater 5(5):1800831
26. Lin C-T, Hsu C-H, Lee C-H, Wu W-J (2011) Inkjet-printed organic field-effect transistor by using composite semiconductor material of carbon nanoparticles and poly(3-Hexylthiophene). J Nanotechnol, 1–7
27. Tan HW, Tran T, Chua CK (2016) A review of printed passive electronic components through fully additive manufacturing methods. Virtual Phys Prototyp 11(4):271–288
28. Muth JT, Vogt DM, Truby RL, Mengüç Y, Kolesky DB, Wood RJ, Lewis JA (2014) Embedded 3D printing of strain sensors within highly stretchable elastomers. Adv Mater 26(36):6307–6312
29. Heckman EM, Bartsch CM, Kreit EB, Aga RS, Ouchen F (2020) Printed electronics for aerospace applications, 93–104
30. Lu Y, Vatani M, Choi J-W (2013) Direct-write/cure conductive polymer nanocomposites for 3D structural electronics. J Mech Sci Technol 27(10):2929–2934
31. Goh GL, Zhang H, Chong TH, Yeong WY (2021) 3D printing of multilayered and multimaterial electronics: a review. Adv Electron Mater 7(10):2100445

Electrochemical Micromachining: A Review on Principles, Processes, and Applications

Rishikant Mishra, Ravi Pratap Singh, and R. K. Garg

1 Introduction

Industries such as automotive, aerospace, electronics, optics, medical devices, and communications have experienced a significant increase in the necessity for large- and small-scale products and parts created from machining-impossible materials including titanium alloys, super alloys, carbides, tool steel, and carbides. Despite having remarkable properties, many of these materials are underutilized. Using traditional machining techniques like turning and milling on these materials is challenging. For example, titanium alloys tend to harden during work and have poor heat conductivity and high chemical reactivity, which leads to high cutting temperatures, tool wear, and strong adherence to the workpieces. A better option for creating precise 3D complicated shapes characteristics and parts out of difficult-to-machine materials is electrochemical micromachining (ECMM). This technique employs electrochemical reactions to selectively remove material from the workpiece, resulting in the fabrication of micro-objects with high precision and accuracy, including complex geometries and high aspect ratio structures. ECMM is a versatile and cost-effective method that can be applied to a wide range of materials, such as metals, semiconductors, and ceramics. Its ability to create microstructures with high precision and accuracy has made electrochemical micromachining a promising option for various applications, such as micro-electromechanical systems (MEMS), microfluidics, and biomedical devices. The utilization of challenging-to-process substances such as titanium alloy and Inconel alloy has notably risen in present times. There is a group of

R. Mishra (✉) · R. K. Garg
Department of Industrial and Production Engineering, Dr B R Ambedkar National Institute of Technology, Jalandhar, Punjab, India
e-mail: rishikantm.ip.22@nitj.ac.in

R. P. Singh
Department of Mechanical Engineering, National Institute of Technology Kurukshetra, Thanesar, Haryana, India

V. S. Sharma et al. (eds.), *Machining and Additive Manufacturing*, Lecture Notes in Mechanical Engineering, https://doi.org/10.1007/978-981-99-6094-1_5

nickel-based superalloys that can uphold their durability despite being repeatedly exposed to elevated temperatures. This alloy possesses improved welding characteristics and has robust strength, high-temperature resistance, and protection against corrosion [1].

There are several variations of the process being researched around the world right now, including pulsed electrochemical machining (PECM), through-mask electrochemical micromachining (TMECMM), wire-electrochemical machining (Wire-ECM), jet-electrochemical machining (Jet-ECM), and electrochemical grinding (ECG). From a fundamental standpoint, all variations of ECM use the same material removal mechanism (anodic dissolution at the ionic level in the presence of electrolyte), but they vary in the size, form, and kinematics of the tool and workpiece [2]. The anodic dissolution rate, surface topology, and precision of the machined component or feature in the ECM process are influenced by the kinematics and elemental composition of the electrolysis in the electrode spacing. During any electrochemical reaction, mass movement significantly affects these parameters [3]. As the electrochemical dissolution process progresses, byproducts of electrolysis, such as air pockets and precipitates begin to form in the interelectrode gap and influence the electrical conductivity of the electrolyte. For better mass transportation, often a speed of (15–45 m/s) electrolyte jet is needed (hence, machining accuracy). Due to the electrolyte's high momentum, electrolysis products are efficiently and swiftly evacuated from the cutting zone as sludge. In the case of micromachining, however, issues including vibrations of the machine's structure and microtools as well as electrolyte loss must be dealt with. An alternative approach to isolate the dissolving region and promote effective sludge draining is to use a pulsating power source. In this scenario, the electrolysis is allowed to relax between two successive pulses by following the machining time with idle time. The alternating current method for various anodic dissolutions of low-carbon steel and copper was the prior name for the pulsing technique [3–5].

Until now, various experimental studies have been conducted on microelectrochemical machining to explore the impact of multiple process parameters on its performance. These parameters can be classified into six categories, as illustrated in Fig. 1 using a fishbone diagram. For the purpose of microelectrochemical machining, to obtain precise and controlled material removal, optimizing various machining settings is crucial. The tool feed rates, electrolyte concentration, electrolyte concentration, applied voltage, and pulse amplitude, all have an impact on how much material is removed from the surface and how rough it is.

The impact of various process parameters on the performance of micro-ECM is outlined in Table 1. This paper aims to bridge a significant gap in the current state of knowledge regarding ECMM and its hybrid variants. This review aims to better understand ECMM with microelectrochemical machining-based hybrid micromachining technologies in a single document because the creation of hybrid processes necessitates a thorough grasp of the parent process. While there are fundamental reviews on electrochemical micromachining in the literature, like those found in references [9–12], these articles only cover the early fundamental understanding of

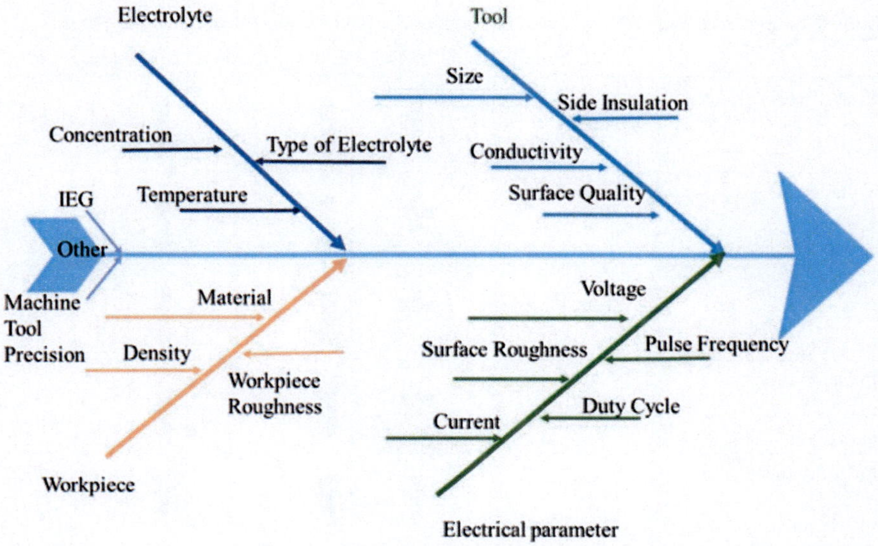

Fig. 1 Fish-bone diagram

the field. References [9] and [10] provided an early introduction to localized electrochemical machining processes many years ago, while Bhattacharya et al. [11] presented the basics of ECMM. Landolt et al. [12] focus on fundamental chemical factors, mass transport effects, passive oxide laser etching, and the formation of multi-level structures. A review of electrochemical machining drilling by Sen et al. [13] focused on hole quality and the effects of process factors. They did not, however, fully address issues like tooling, interelectrode spacing, energy sources, and the choice of electrolyte for particular materials. It should be noted that since the review was conducted, technology has rapidly progressed. A conference review article [14] provided a brief overview of ECM technology and did not go into the specifics of the processes involved. It mainly consisted of information gathered from existing literature on ECM. On the other hand, Leese et al.'s work [15] presented an overview of process parameters involved in ECM. In this review article provided, an overview of ECM technology by including the latest research from both academic and industrial fields. This article is aimed at audiences from both academia and industry.

2 Principle of Material Removal

The fundamental idea behind the removal of material in microelectrochemical machining (micro-ECM), which involves anodic dissolution of the workpiece, is the same as that in electrochemical machining. In contrast, the goal of microelectrochemical machining is to pinpoint material removal in order to obtain exact control

Table 1 Observed parameter of process performance in μ-electrochemical machining [6–8]

	Machining gap	MRR	Surface finish	Shape precision
Electrolyte concentration	+	+	↓	↓
Voltage	+	+	↓	↓
Duty cycle	+	+	↓	↓
Pulse on time	+	+	↓	↓
Pulse frequency	+	+	↓	↓

over the morphology. Compared to other localized material removal procedures, micro-ECM provides a number of benefits. For instance, because the dissolving takes place at the atomic level, it is unaffected by the workpiece's hardness, can be utilized to make complex shapes, and has no tool wear. As a result, the surface quality is very good.

Micro-ECM may be precisely controlled in micromachining by applying ultra-short pulse power. Furthermore, it is a non-contact technique that doesn't generate any machining forces, making size effects irrelevant. To improve process capabilities and widen the material processing area, micro-ECM can potentially be integrated with several other processes. Finally, electrical factors like electrical current and voltage as well as pulse properties like frequency ratio, on time, length, and pulse duration can be used to accurately control material removal rates in micro-ECM. The workpiece serves as the anode as well as the tools electrode as the cathode in an anodic reaction. The metallic workpiece goes through oxidation during this process, which releases electrons [16].

Various process inputs utilized in the electrochemical machining process are shown in Fig. 1 with the help of the Ishikawa diagram in which four main factors, i.e., electrolyte, tool used in the process, workpiece nature, and electrical parameters considered in the process are highlighted with their subparts.

Schematic diagram of ECM process with the pulse power supply is shown in Fig. 2. The main elements of the ECM process are connected in series.

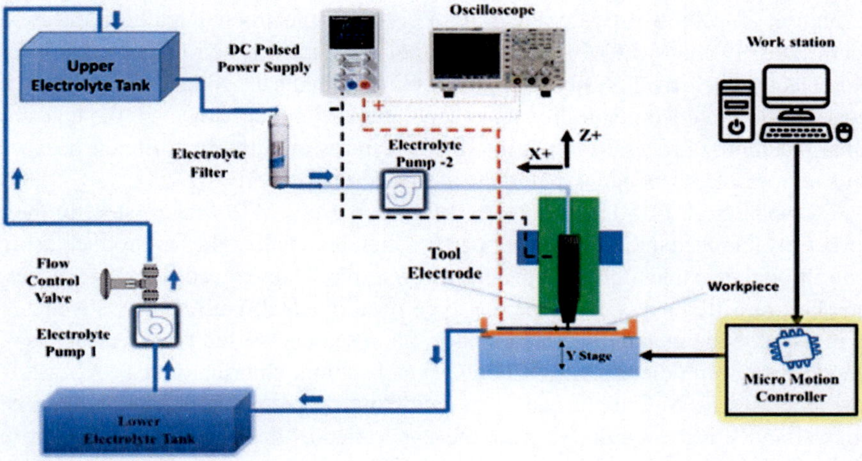

Fig. 2 Schematic diagram of ECMM

3 Literature Review

Kumar et al. [17] conducted experiments on the floating metal powder with moving on Stainless steel 316L through the ECMM process. The purpose of the experiment was to investigate the effectiveness of ECMM in cutting SS316L using citric acid as an environmentally friendly electrolyte. The results of employing copper powder material immersed in a citric acid electrolyte without the aid of a mechanical overhead stirrer were compared to those of using copper powdered material immersed in an acid electrolyte with that assistance. To determine the output performance characteristics, such as material removal rate and overcut, the researchers changed the input effect of various parameters, including operating voltage, pulse duration, and solution concentration.

A study conducted by Ayyappan et al. [18] investigate the use of a magnetic flux-assisted low-frequency vibrating tool as a productive hybrid ECM method to improve MRR and surface quality index. The researchers created a mathematical model that connected surface roughness with machining factors like voltages, solution concentration, and IEG. Despite its potential benefits, ECM was not widely used due to accuracy issues, challenges in tool design, and parameter control difficulties. Using contour plots, the importance of electrochemical machining process variables was examined.

Singh et al. [19] introduced a new experimental technique, known as "one parameter at a time" approach, to evaluate a diamond cut-off grinding system using electrochemical methods on a locally made setup. The study focused on the grinding of nickel-based superalloy (Inconel 925), which is challenging to machine using traditional methods. Electrochemical grinding is a valuable technique for machining conductive materials that are newly developed or difficult to cut.

Sethi et al. [20] investigate how nitinol dissolves electrochemically in different electrolytes for micro-ECM purposes. Nitinol is an ideal shape memory alloy for microelectromechanical systems, especially in biomedical applications, due to its super elasticity, biocompatibility, and shape memory effect. However, the conventional machining process for creating micro features on nitinol is difficult because of its temperature-dependent material transformation properties.

Ayyappan et al. [21] investigate the ECM properties of 20MnCr5 steel in their work. Certain mechanical regulating parts, which are challenging to produce using conventional machine tools, like pistons screws, spindle, overhead cams, gearboxes, and shaft, are frequently made of this type of alloy steel, particularly, 20MnCr5. To investigate the machining performance, the researchers used two electrolytes—aqueous potassium dichromate (K2Cr2O7) and sodium chloride (NaCl). K2Cr2O7 was added to the NaCl bath due to its oxidizing properties. The rate of material removal and surface roughness were the two variables that the study looked at in relation to common electrochemical machining parameters such voltage level, IEG, and solution concentration.

Cao et al. [22] study to investigate the behavior of TA15 anode in anode dissolving and obtain the necessary surface attributes for counter-rotating electrochemical machining. The researchers measured the anodic characteristics and examined the passive–trans passive behavior of TA15 using a polarized and voltammetry curve. Unfortunately, limited research has been done on the anodic behavior of the counter-rotating state. A quantitative dissolution model was developed to explain the electrochemical dissolution and structural evolution of the revolving surfaces after electrochemical impedance spectroscopy was used to investigate the electrode structure at various phases.

Rahi et al. [23] conducted a study to use an electrochemical method to cut metal matrix composites. However, they faced two significant obstacles: low material removal rate and oxide layer buildup on the machined surface. To overcome these issues, they proposed a hybrid approach combining conventional grinding and electrochemical machining. To test this approach, they designed an electrochemical surface grinding experiment to compare its effectiveness with the ECM method for cutting the challenging Al-SiC-Gr metal matrix composite.

Rajesh et al. [24] examined how several electrochemical hole-digging process variables, such as supply voltage, electrolyte solution, and pulse duration, affected the rate at which material was removed from aluminum 7075 alloy composites containing silicon carbide and fly ash particles. These composites have potential applications in the construction of trusses, frameworks, and vessels that hold pollutants, dairy, and other acidic substances in areas with high salt concentration, as an alternative to stainless steel, which is commonly used to prevent corrosion.

The impact of several factors on the macroelectrochemical machining of aluminum alloy 7075 composite with silicon carbide and fly-ash was examined by Chinailli et al. [25]. To examine the effects of process variables such voltages, solution concentration, interelectrode spacing, as well as drilling time, the researchers employed a central composite design technique. According to the study, drilling time, voltage, and electrolyte concentration all had a substantial impact on the drilled hole removal rate of material and surface quality. The surface morphology and elemental makeup of the drill holes were also examined by the researchers using scanning electron microscopy (SEM) and energy-dispersive X-ray spectroscopy (EDX).

Singh and Sharma [26] investigated the usage of innovative thermoplastic slit profiles in electrochemical machining (ECM). The performance of the thermoplastic slit profiles in comparison to traditional slit profiles was assessed using a micro-ECM setup in terms of the amount of material removed and surface quality. The study found that the thermoplastic slit profiles produced a higher MRR and a lower SR compared to the conventional slit profiles. The researchers also observed a longer tool life for the thermoplastic slit profiles, indicating that they had less tool wear compared to the conventional slit profiles.

Vijayakumar et al. [27] studied the optimizing electrochemical micromachining (ECMM) process for a copper alloy. The removal rate of material and surface quality are two examples of the output parameters that the authors investigate using a (RSM). Voltage, solution concentration, and IEG are the three input parameters that are varied in the experimental design. The authors use a Box-Behnken design to generate 15 experimental runs, with the removal rate of material and surface quality observed from each run. The resulting data is used to build regression models for each output parameter, which are then optimized using a multiobjective genetic algorithm.

Aravind and Hiremath [28] developed a model to optimize the microelectrochemical machining (micro-ECM) parameters for machining holes on copper plates. The impact of various input variables, such as ion concentration, voltage, and machining time, on the rate of material removal and surface morphology of the machined holes was examined by the researchers using a mathematical model and the response surface methodology (RSM). The surface shape and elemental makeup of the machined surface were further examined using an EDS and a scanning electron microscopy. The findings demonstrated that a greater MRR and a lower SR were produced by the ideal set of input parameters.

Kumar et al. [29] investigated the use of electrochemical jet machining (ECJM) for surface finishing of an additively manufactured (AM) part. The researchers used a stainless steel AM part and evaluated the surface roughness (SR) and material removal rate (MRR) of the part using ECJM. To examine the impact of various parameters, including electrolyte concentration, tools feed rate, and applied voltage on the SR and MRR, the study employed a Taguchi-based methodology.

In order to increase the removal rate of material (MRR) and decrease sludge formation during the ECM of the Monel 400 alloy, Nagarajan et al. [30] examined the application of a meta-heuristic approach. To examine the impact of multiple input factors, including machining voltage, solution concentration, tools feed rate, and machining time on the MRR and sludge formation, the researchers employed a Taguchi-based design of experiments (DOE) strategy. The grey wolf optimization (GWO) algorithm was employed in the study to enhance the input variables and improve the efficiency of the machining. The results showed that the GWO algorithm produced better results compared to the Taguchi-based DOE approach in terms of MRR and sludge reduction.

Deepak and Hariharan [31] investigated the influence of auxiliary electrodes and magnets on electrochemical machining (ECM) of SS304 using NaCl and NaNO3 electrolytes. The experimental setup involved a stainless steel workpiece, a copper auxiliary electrode, and a magnet, which were placed in the electrolyte solution. The outcomes showed that the addition of copper auxiliary electrodes reduced the surface roughness of the machined surface, with a smoother surface being produced by the NaNO3 electrolyte than by NaCl. The presence of a magnet had a considerable impact on the rate of machining, with the magnetized setting outperforming the non-magnetized setup.

Kumar et al. [32] investigate the impact of tool rotation on the fabrication process of micro-tools using electrochemical micromachining (ECMM) technology. ECMM is a non-conventional process that is used to fabricate micro- and nano-scale features on conductive materials. The study evaluates how the ECMM process is affected by two alternative tool rotation speeds. The researchers have found that a fast-rotating tool produced a more even distribution of material removal throughout the tool surface, producing a smoother tool surface and a more exact shape. They also found that higher rotation speeds allowed for a faster fabrication process, with a higher material removal rate.

Sahai et al. [33] investigate a new method for micromachining silicon using a combination of electrochemical and spark-assisted milling techniques. The study aims to develop an efficient and cost-effective method for fabricating microstructures on silicon, which is an important material for microelectromechanical systems (MEMS) and microfluidics. The researchers first developed a mathematical model of the M-ECSMM process, which takes into account the various parameters involved in the machining process, such as voltage, current, and electrolyte concentration.

4 Future Scope and Research Aspects for Research in ECMM

Electrochemical micromachining (ECMM) is a specialized field that involves the use of electrochemical processes to fabricate miniature structures and devices with high precision and accuracy. The technology has numerous applications across several industries, including microelectronics, medical devices, and aerospace. In recent years, there has been a growing interest in the use of ECMM for fabricating micro-electromechanical systems (MEMS) and microfluidic devices. These tiny devices have the potential to revolutionize several fields, including biomedical research, drug delivery, and diagnostic testing. The future scope of research in ECMM is quite promising, and some of the key areas of focus include:

1. Development of new electrochemical machining techniques: Researchers are constantly working to improve existing ECMM techniques and develop new methods that can produce even more precise and complex structures.

2. Advancement of MEMS technology: The development of MEMS technology is a major focus in ECMM research, and future advancements in this area could lead to the creation of new devices with improved capabilities and functionality.
3. Integration of ECMM with other fabrication techniques: The integration of ECMM with other fabrication techniques such as photolithography and 3D printing could lead to the creation of even more complex structures and devices.
4. Applications in biomedical research: ECMM has several potential applications in biomedical research, including the fabrication of microscale sensors for monitoring biological processes and the creation of microfluidic devices for drug delivery.

Overall, the future scope of research in ECMM is quite broad, and continued advancements in this field could have significant impacts across several industries.

5 Conclusion

Based on a study of numerous machining methods, researchers have concluded that ECMM is highly useful for machining aero-engine components, biomedical components, etc. This review article proposes a potential advancement in electrochemical machining or hybrid techniques, as well as the design of intricate machining profiles, numerical analysis of the electrochemical process, and prediction of anode shape using flow field simulation. The following recommendations are made for additional research.

1. Precision and necessary geometric accuracies can be challenging to achieve when machining high-strength material. Electrochemical micromachining (ECMM), one of the available machining techniques, is frequently used to make products of excellent quality from metals and alloys and can minimize machining time and expense. According to the available literature, it has been shown that employing the electrochemical micromachining technique to machine these high-temperature titanium alloys produced certain results that are appealing to users in the industry and provide room for future research.
2. The majority of research is focused on fundamental questions, such as how well different materials can be machined, how well processes can be developed, how to use pulsed power supplies, and how to use parametric studies to evaluate process performance. There are a few areas that still require investigation and are not fully developed: development of specialized machinery, evaluation of the micro-ECM process's product and process characteristics, the correlation between material clearance and micro-ECM pulses, interelectrode phenomena characterization using customized settings, multidisciplinary modeling, which takes into account bubble phenomena, mass and charge movement, and the fundamental concept of the micro-ECM process creation of ecologically friendly electrolytes and alternatives for acidic electrolytes.

3. The design and development of universal or multifunctional machine tools that can be used with multiple process energies, understanding of material removal mechanisms on different materials when two or more process energies act simultaneously in the same machining zone, and synchronization of two process energies on the same machining axis to control precision of material removal and shape control are a few of the critical elements that literature has not yet fully addressed. enhancing knowledge of the interplay between two process energy.
4. ECMM offers several advantages over traditional micromachining techniques, such as high accuracy, excellent surface finish, and the ability to machine complex geometries. ECMM can be used for a variety of applications, including micro-electrodes, micromolds, microfluidic devices, and micro-sensors. The process capabilities of ECMM can be enhanced by combining it with other micromachining technologies, such as micro-milling, microgrinding, and laser machining, to form hybrid variants.
5. The development of micro-scale features on a variety of materials, including metals, semiconductors, and insulators, is made possible by the promising ECMM technology. Hybrid variants of ECMM can provide additional benefits, such as improved material removal rate, reduced machining time, and enhanced surface quality.
6. The development of a passivation layer on the workpiece surface, which can restrict the rate of material removal and lower machining precision, is one of the main difficulties faced by ECMM. Several approaches have been developed to overcome this challenge, including the use of high-frequency pulsing, the application of ultrasonic vibration, and the addition of surface-active agents.
7. The choice of electrolyte solution is critical for achieving high-quality machining results in ECMM. The electrolyte solution should have high conductivity, low viscosity, and high stability. The process variables, like feed rate, voltages, solution concentration, and current density, greatly affect how well the ECMM can machine materials.

Overall, ECMM and its hybrid variants offer great potential for producing high-quality microscale features on a wide range of materials, and further research and development are needed to fully explore their capabilities and applications.

Acknowledgements The authors would like to recognize the monetary support attained from SERB, New Delhi, India under grant no EEQ/2021/000031.

References

1. Sharman ARC, Hughes JI, Ridgway K (2004) Workpiece surface integrity and tool life issues when turning inconel 718TM nickel based superalloy. Mach Sci Technol 8(3):399–414. https://doi.org/10.1081/MST-200039865
2. Patel D, Jain VK, Ramkumar J (2018) Micro texturing on metallic surfaces: state of the art. Proc Inst Mech Eng Part B J Eng Manuf 232(6): 941–964

3. Datta M, Landolt D (1975) Surface brightening during high rate nickel dissolution in nitrate electrolytes. J Electrochem Soc 122(11):1466–1472. https://doi.org/10.1149/1.2134043
4. Berkeley L (2011) This is a library circulating copy
5. Mao K-W (1973) Anodic polarization study of mild steel in NaCl solution during electrochemical machining. J Electrochem Soc 120(8):1056. https://doi.org/10.1149/1.2403627
6. Bhattacharyya B, Munda J (2003) Experimental investigation on the influence of electrochemical machining parameters on machining rate and accuracy in micromachining domain. Int J Mach Tools Manuf 43(13):1301–1310. https://doi.org/10.1016/S0890-6955(03)00161-5
7. Kim BH, Ryu SH, Choi DK, Chu CN (2004) Micro electrochemical milling. J Micromechanics Microengineering 15(1):124
8. Kurita T, Chikamori K, Kubota S, Hattori M (2006) A study of three-dimensional shape machining with an ECμM system. Int J Mach Tools Manuf 46(12–13):1311–1318. https://doi.org/10.1016/j.ijmachtools.2005.10.013
9. Van Osenbruggen C (1985) Electrochemical micromachining. Philips Tech Rev 42:22
10. Papers K (1999) New developments in electro-chemical machining. 48(2):567–579
11. Bhattacharyya B, Munda J, Malapati M (2004) Advancement in electrochemical micromachining. Int J Mach Tools Manuf 44(15):1577–1589. https://doi.org/10.1016/j.ijmachtools.2004.06.006
12. Landolt D, Chauvy PF, Zinger O (2003) Electrochemical micromachining, polishing and surface structuring of metals: fundamental aspects and new developments. Electrochim Acta 48(20–22):3185–3201. https://doi.org/10.1016/S0013-4686(03)00368-2
13. Sen M, Shan HS (2005) A review of electrochemical macro-to micro-hole drilling processes. Int J Mach Tools Manuf 45(2):137–152. https://doi.org/10.1016/j.ijmachtools.2004.08.005
14. Rajurkar KP, Sundaram MM, Malshe AP (2013) Review of electrochemical and electrodischarge machining. Procedia CIRP 6:13–26. https://doi.org/10.1016/j.procir.2013.03.002
15. Leese R, Ivanov A (2018) Electrochemical micromachining: review of factors affecting the process applicability in micro-manufacturing. Proc Inst Mech Eng Part B J Eng Manuf 232(2):195–207. https://doi.org/10.1177/0954405416640172
16. Joshi SS, Marla D (2014) 11.15-electrochemical micromachining. Compr Mater Process, 373–403. Oxford Elsevier
17. Vinod Kumaar JR, Thanigaivelan R, Soundarrajan M (2022) A performance study of electrochemical micro-machining on SS 316L using suspended copper metal powder along with stirring effect. Mater Manuf Process 37(13):1526–1539. https://doi.org/10.1080/10426914.2022.2030874
18. Ayyappan S, Sivakumar K, Kalaimathi M (2017) Electrochemical machining of 20MnCr5 alloy steel with magnetic flux assisted vibrating tool. Proc Inst Mech Eng Part C J Mech Eng Sci 231(10):1956–1965. https://doi.org/10.1177/0954406215623310
19. Singh R, Narayan A (2022) Parametric study of indigenously developed electrochemical diamond cut-off grinding setup using inconel 925. Int J Mach Mach Mater 24(5):370–390
20. Sethi A, Acharya BR, Saha P (2022) Study of the electrochemical dissolution behavior of nitinol shape memory alloy in different electrolytes for micro-ECM process. Int J Adv Manuf Technol 121(9–10):7019–7035. https://doi.org/10.1007/s00170-022-09802-z
21. Ayyappan S, Sivakumar K (2015) Investigation of electrochemical machining characteristics of 20MnCr5 alloy steel using potassium dichromate mixed aqueous NaCl electrolyte and optimization of process parameters. Proc Inst Mech Eng Part B J Eng Manuf 229(11):1984–1996. https://doi.org/10.1177/0954405414542136
22. Cao WJ, Wang DY, Cui GW, Zhu D (2022) Anodic dissolution mechanism of TA15 titanium alloy during counter-rotating electrochemical machining. Sci China Technol Sci 65(6):1253–1262. https://doi.org/10.1007/s11431-021-1999-7
23. Rahi DK, Dubey AK (2022) Comparative study of machining quality for the Al--SiC--Gr hybrid metal matrix composite using ECM and ECSG. Proc Inst Mech Eng Part C J Mech Eng Sci 236(12):6740–6751
24. Rajesh S et al (2022) Electrochemical machining of aluminium 7075 alloy, silicon carbide, and fly ash composites: an experimental investigation of the effects of variables on material removal rate. Mater Today Proc 62:863–867. https://doi.org/10.1016/j.matpr.2022.04.054

25. Swamy Chinamilli NVS et al (2023) An empirical study of the impacts of parameters in micro electrochemical drilling of Al 7075 alloy composites with SiC and fly-ash. Mater Today Proc. 72:2464–2468. https://doi.org/10.1016/j.matpr.2022.09.453
26. Singh R, Sharma AD (2023) Innovative thermoplastic slit profiles for electrochemical machining. Int J Interact Des Manuf. https://doi.org/10.1007/s12008-022-01174-8
27. Vijayakumar K, Jayaseelan GAC, Shakil OV M, Asharaf MJ M, Nafeel VC M (2023) Multi-objective optimization of electrochemical micromachining of copper alloy by response surface methodology. AIP Conf Proc 2523(1):20001
28. Aravind S, Hiremath SS (2023) Modelling of micro-electrochemical machining parameters used for machining of holes on copper plate. J Indian Chem Soc 100(3):100933. https://doi.org/10.1016/j.jics.2023.100933
29. Kumar P, Dixit P, Chaudhary B, Jain NK (2023) Surface finishing of an additively manufactured part using electrochemical jet machining. Mater Today Commun 35 December 2022, 105581. https://doi.org/10.1016/j.mtcomm.2023.105581
30. Nagarajan V et al (2022) Meta-heuristic technique-based parametric optimization for electrochemical machining of monel 400 alloys to investigate the material removal rate and the sludge. Appl Sci 12(6). https://doi.org/10.3390/app12062793
31. Deepak J, Hariharan P (2022) Study on the influence of auxiliary electrode and magnet on electrochemical machining of SS304 using NaCl and NaNO3. Surf Topogr Metrol Prop 10(1):15045
32. Kumar A, Yadav HNS, Kumar M, Das M (2022) Effect of tool rotation on the fabrication of micro-tool by electrochemical micromachining. J. Micromanufacturing 5(2):217–223. https://doi.org/10.1177/25165984211031687
33. Sahai K, Narayan A, Yadava V (2023) Development and experimental study of milling electrochemical spark micromachining (M-ECSMM) of silicon. SILICON 15(1):473–497. https://doi.org/10.1007/s12633-022-02023-9

Friction and Wear Characteristics of Engine Oil Through Four-Ball Tester

Venkateswara Babu Peddakondigalla, Ismail Syed, Vasavi Boggarapu, P. S. Rama Sreekanth, Santosh Kumar Sahu, Venkata Dinesh Avvari, Rajani Doppalapudi, and Lakshmi Sunitha Vikram

1 Introduction

It is well-known that proper lubrication is one of the critical factors that enhance the tribological behavior of interacting sliding surfaces. Lubricant is a widely used element between the interacting surfaces in various mechanical components, which are in relative motion. The condition and performance of the lubricant oil affect the overall performance of the mechanical systems. The properties of the engine oil reflect the performance of an internal combustion (IC) engine, and thus affect fuel consumption [1]. The primary purpose of a lubricating oil is to enhance the anti-friction and anti-wear behavior of the interacting surfaces by avoiding metal-to-metal contact. In addition, it can also reduce corrosion, temperature, contamination, and vibrations between the sliding surfaces. Proper lubrication enhances the working life as well as the fuel efficiency of the engine. The tribological properties such as friction and wear of engine components can significantly affect the engine performance. If the tribological performance of the engine components decreases, the working life of the engine reduces. The relationship between lubricating oil and the performance

V. B. Peddakondigalla
Department of Mechanical Engineering, Vasireddy Venkatadri Institute of Technology, Guntur, Andhra Pradesh, India

I. Syed
Department of Mechanical Engineering, National Institute of Technology Warangal, Hanamkonda, Telangana, India

V. Boggarapu · P. S. R. Sreekanth (✉) · S. K. Sahu · V. D. Avvari
School of Mechanical Engineering, VIT AP University, Amaravati, Andhra Pradesh, India
e-mail: happyshrikanth@gmail.com

R. Doppalapudi
Department of English, RVR and JC College of Engineering, Guntur, Andhra Pradesh, India

L. S. Vikram
Department of Physics, RVR and JC College of Engineering, Guntur, Andhra Pradesh, India

V. S. Sharma et al. (eds.), *Machining and Additive Manufacturing*, Lecture Notes in Mechanical Engineering, https://doi.org/10.1007/978-981-99-6094-1_6

Table 1 Properties of selected lubricants

Lubricant	Density (gm/cm^3)	Flash point (°C)	Kinematic viscosity at 40°C (cSt)
SAE20W-50	0.87	214	172.3
SAE15W-40	0.866	232	107
SAE10W-30	0.875	210	67.5

of the engine is an important diagnostic tool to decide the overall efficiency of the engine [2].

Recently, some of the researchers used the synthetic engine oils such as SAE15W-40 and SAE10W-30 for tribological analysis of sliding pairs under different contact conditions [3–7]. Furthermore, many researchers utilized the four-ball tester assembly in order to evaluate the tribological performance, load-carrying capacity, extreme pressure, and extreme temperature analysis of the different lubricants under different scenarios [8–13]. In the present work, different engine oils are tested in order to select the appropriate lubricant for better tribological performance. The experiments are conducted on a four-ball tester and the lubrication performance of considered engine oils is characterized.

2 Experiment

2.1 *Test Lubricants*

In the present work, three lubricants, namely, SAE20W-50, SAE15W-40, and SAE10W-30, were selected, which are commercially available lubricants for automobiles. The selected lubricants are synthetic in nature, and the viscosity of these lubricants does not change much during the test period. Castrol India Limited supplies all the selected lubricants, and the same brand has been used during all the tests. These are tested on a four-ball tester for a better understanding of their frictional and wear properties. The properties of the selected lubricants are presented in Table 1.

2.2 *Apparatus*

The selected lubricants are tested for better lubricity performance in terms of friction and wear properties. The friction and wear tests are conducted on a four-ball tester (Ducom made, model: TR-30L-IAS) according to a standard test procedure of ASTM D4172. The schematic view of the four-ball tester assembly is shown in Fig. 1. The four-ball tester assembly consists of four steel balls out of which three balls remain stationary, while the fourth ball rotates against these three balls (see Fig. 1).

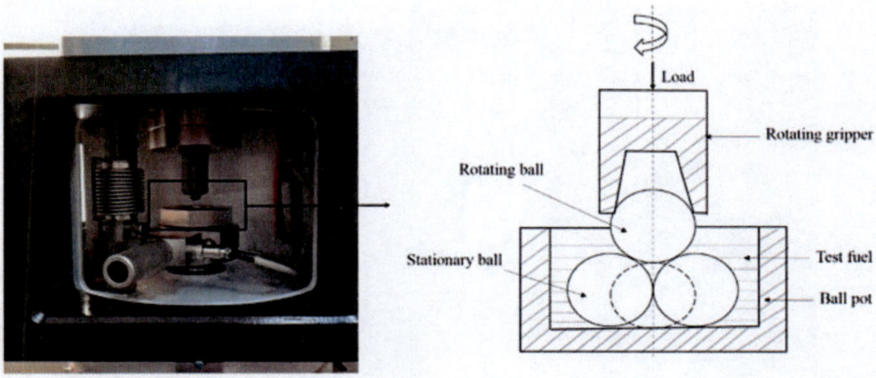

Fig. 1 Four-ball tester assembly

Before each test, the four steel balls are cleaned with acetone and wiped dry with a tissue. Three steel balls are placed in a ball pot and tightened to a torque of a minimum 67 Nm using a torque wrench in order to avoid relative motion between them. Then, 10 ml of test lubricant is poured into the ball pot and ensured that the lubricant fills all the voids in the ball pot. This ball pot is placed beneath the spindle to which the fourth ball is fixed through a collet chuck, and this ball rotates against three balls, which are placed in the ball pot.

The operating conditions of a four-ball tester are shown in Table 2. Each test is conducted for 1 h, and new steel balls are used for each test. After each test, the frictional torque can be measured with the help of a torque sensor for the corresponding applied normal load. Afterward, the friction coefficient is calculated by Eq. (1).

$$Friction\ coefficient = \frac{T\sqrt{6}}{3W_1 r} \quad (1)$$

Table 2 Operating conditions of four-ball tester

Parameter	Condition
Experimental conditions (ASTM D4172)	
Load	15 kg
Speed	600, 900, 1200, 1500 rpm
Test duration	1 h
Testing ball conditions	
Ball material	AISI E-52100
Ball diameter	12.7 mm
Ball hardness	62 HRC
Ball roughness	0.1 μm

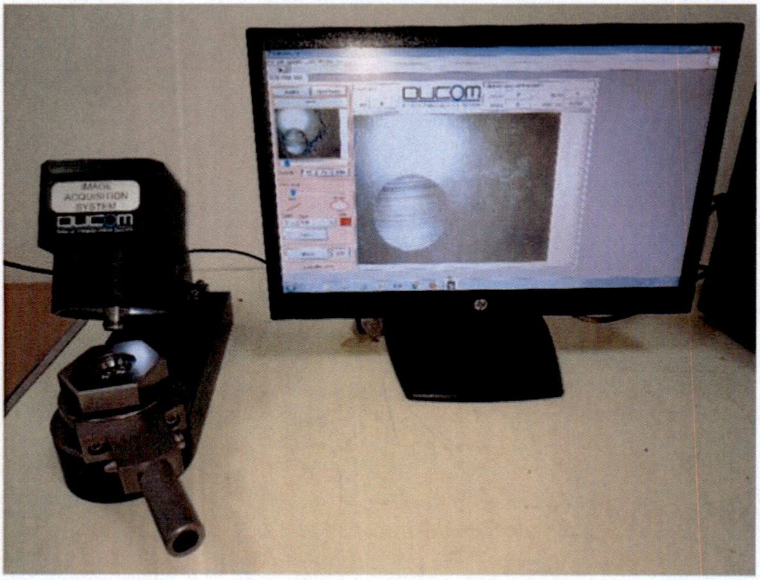

Fig. 2 Wear scar captured by image acquisition system

where, T is the frictional torque in Kg-mm, W_1 is the applied load in Kg, and r (3.67 mm) is the center distance between the contacting lower ball surfaces to the axis of rotation.

After the completion of each test on the four-ball tester, the wear scar diameter of three stationary balls which are in ball pot is measured with the help of an image acquisition system, as shown in Fig. 2. It consists of a camera to capture the image of the wear scar and a compound lens to capture the image at predefined focal length to measure the wear scar diameter. The wear scar can be viewed on the PC with the help of SCARVIEW software, and the wear scar is measured using the ellipse tool. The focal length can be adjusted to get a clear image, and this image is captured by the camera, which would be transferred and stored in a PC.

3 Results and Discussion

The variation of frictional torque for different test lubricants is depicted in Fig. 3. The tests are conducted at an applied load and rotating speed of 15 kg and 1200 rpm, respectively. Among the tested lubricants, SAE10W-30 shows maximum frictional torque while SAE15W-40 exhibits minimum frictional torque (see Fig. 3).

Figure 4 shows the measured mean friction coefficient and wear scar diameter of different tested lubricants at an applied load of 15 kg and speed of 1200 rpm. It is observed that the mean friction coefficient of SAE20W-50, SAE15W-40, and

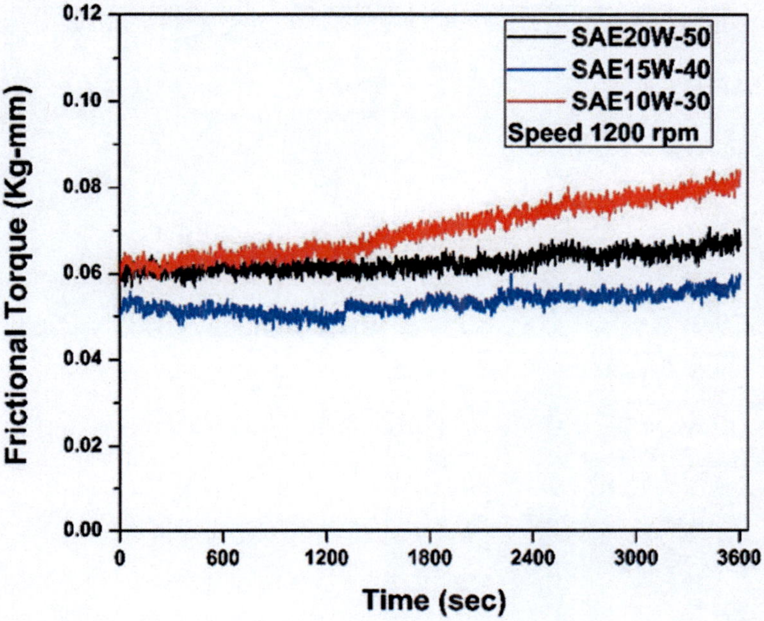

Fig. 3 Frictional torque versus time for different lubricants

SAE10W-30 lubricants are 0.097, 0.092, and 0.116, respectively (see Fig. 4a). The mean wear scar diameter for SAE20W-50, SAE15W-40, and SAE10W-30 lubricants are 72 μm, 68 μm, and 78 μm, respectively (as can be seen in Figs. 4b and 5). The test results revealed that lubricant SAE15W-40 exhibits excellent frictional and wear resistance when compared with other lubricants. This friction and wear reduction with SAE15W-40 lubricant may be due to lower frictional torque when compared with other lubricants, which can be observed in Fig. 3.

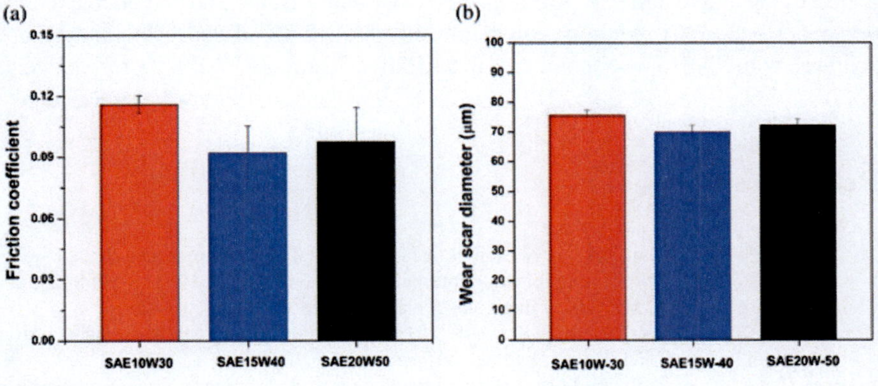

Fig. 4 **a** Friction coefficient and **b** Wear scar diameter of different lubricants

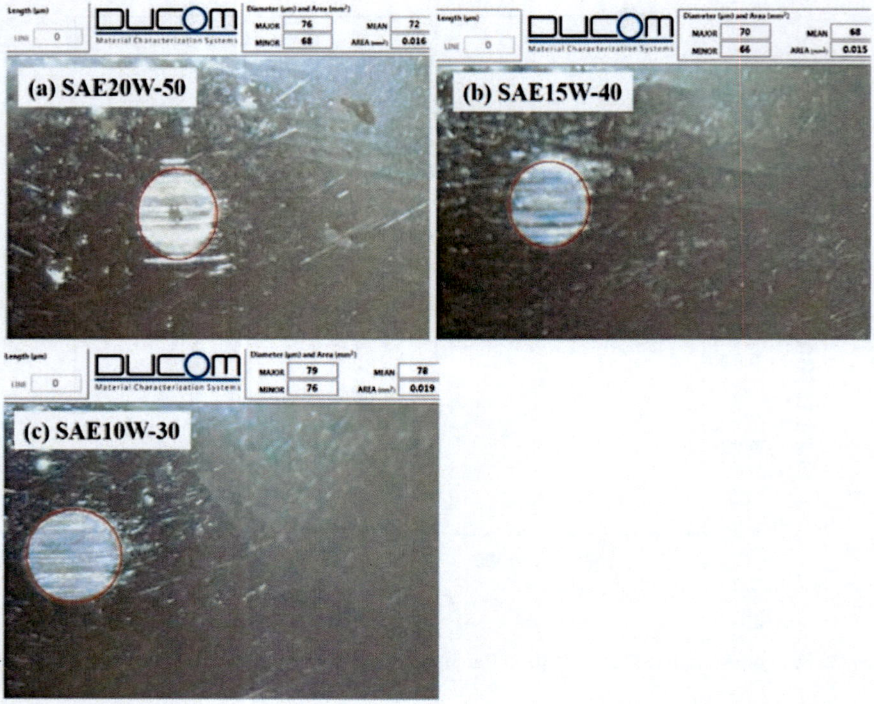

Fig. 5 Measured wear scar diameters of different lubricants by the image acquisition system

4 Conclusion

In the present work, the lubrication performance of various lubricants is tested according to ASTM D4172. The lubrication performance is evaluated in terms of friction coefficient and wear scar diameter. The results exhibited that SAE15W-40 has excellent anti-friction and anti-wear properties, and thus good lubrication performance compared to SAE20W-50 and SAE10W-30 lubricants. The percentage reduction of 5.2%, 20.7% in friction coefficient, and 5.5%, 12.8% in wear scar diameter are achieved with SAE15W-40 compared to SAE20W-50 and SAE10W-30, respectively.

References

1. Yadav G, Tiwari S, Jain ML (2018) Tribological analysis of extreme pressure and anti-wear properties of engine lubricating oil using four ball tester. Mater Today: Proc 5(1):248–253
2. Farhanah AN, Bahak MZ (2015) Engine oil wear resistance. J Tribol 4, 10−20
3. Grabon W, Koszela W, Pawlus P, Ochwat S (2013) Improving tribological behaviour of piston ring–cylinder liner frictional pair by liner surface texturing. Tribol Int 61:102–108
4. Ryk G, Kligerman Y, Etsion I (2012) Experimental investigation of laser surface texturing for reciprocating automotive components. Tribol Trans 45(4):444–449

5. Babu PV, Ismail S, Ben BS (2020) Experimental and numerical studies of positive texture effect on friction reduction of sliding contact under mixed lubrication. Proc Inst Mech Eng, Part J: J Eng Tribol 1350650120930911
6. Babu PV, Syed I, Beera SB (2020) Influence of positive texturing on friction and wear properties of piston ring-cylinder liner tribo pair under lubricated conditions. Ind Lubr Tribol 71(4):515–524
7. Babu PV, Syed I, Ben BS (2020) Optimization of texture geometry for enhanced tribological performance in piston ring-cylinder liner contact under pure hydrodynamic and mixed lubrication. In: Innovative product design and intelligent manufacturing systems. Springer, Singapore, pp 799–808
8. Suresha B, Hemanth G, Rakesh A, Adarsh KM (2020) Tribological behaviour of neem oil with and without graphene nanoplatelets using four-ball tester. Adv Tribol 2020:1–11
9. Aiman Y, Syahrullail S (2017) Development of palm oil blended with semi synthetic oil as a lubricant using four-ball tribotester. J Tribol 13:1–20
10. Saravanakumar N, Saravanan MJ, Barathkumar KE, Kannan KG, Karthikeyan R (2019) Development and testing of nano particulate lubricant for worm gear application. J Mech Sci Technol 33(4):1785–1791
11. Dandan MA, Samion S, Yahaya WMAW, Hadi F (2019) Tribological analysis of modified RBD palm kernel containing anti-oxidant additive using four-ball tribotester. J Adv Res Fluid Mech Therm Sci 56(2):157–164
12. Nazare MI, Paleu V, Bhaumik S, Ianuş G, Olaru DN (2018) Performances of automotive lubricants–tests on four ball machine. MS&E 444(2):022013
13. Deri MM (2019) Evaluation of engine oil mix with palm oil using four ball tribotester with various speed. J Biolubricant Eng Sci 1(2):1–3

Recent Trends in 4th Industrial Revolution for A Sustainable Future–A Review

Suman Gothwal, Alok Vardhan, Ashiwani Kumar, and Pradeep Jain

1 Introduction

Recent technology advances have led to "Industry 4.0". It surpasses mechanisation, industrialisation, and computerisation. Industry 4.0 covers data management, manufacturing processes, competitiveness, and efficiency. Industry 4.0 concept involves information and communication technologies, cyber-physical systems, artificial intelligence, big data analytics, IoT, autonomous robotics, cloud computing, and augmented reality. These technologies may be the main drivers of automated and digital industrial environments. These technologies enable intelligent manufacturing by allowing devices, machines, production modules, and products to autonomously share information, trigger action, and control each other [1]. Industry 4.0 also goals to integrate human personnels into industrial processes to enhance, add value, and reduce waste.

The steam engine sparked the first Industrial Revolution in England in the mid-eighteenth century. Steam and water power mechanised manufacturing [2]. Europe and the United States began the second Industrial Revolution in the late nineteenth century, harnessing electrical power for mass production [3]. The third Industrial Revolution employed electronics and computer technology to automate production in several industrialised countries in the late twentieth century [4, 5]. The steam engine, electricity, and digital technologies used in the first three Industrial Revolutions increased production and efficiency [6]. Industry 4.0's smart factory technologies

S. Gothwal (✉) · A. Vardhan · P. Jain
Department of Mechanical Engineering, Ajay Kumar Garg Engineering College, Ghaziabad 201009, India
e-mail: gothwalsuman@akgec.ac.in

A. Kumar
Department of Mechanical Engineering, Feroze Gandhi Institute of Engineering and Technology, Raebareli (UP), India

V. S. Sharma et al. (eds.), *Machining and Additive Manufacturing*, Lecture Notes in Mechanical Engineering, https://doi.org/10.1007/978-981-99-6094-1_7

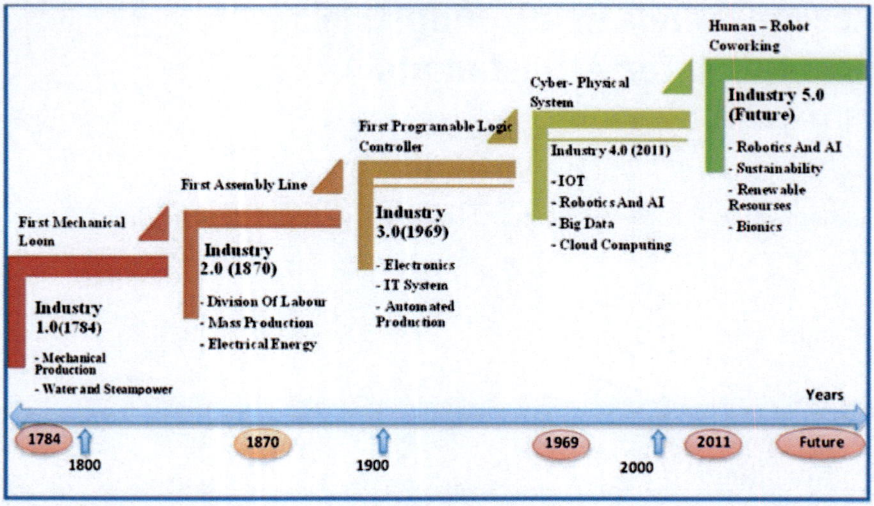

Fig. 1 From industry 1.0 to industry 5.0

have a huge impact on the industry. Figure 1 shows Industrial Revolution 1 to 5 (future) with special features and the year of revolution.

Germany's strategic plan coined "Industry 4.0". Hannover Messe, a German industrial expo, introduced Industry 4.0 in 2011. German Trade and Invest (GTAI) calls Industry 4.0 "a paradigm change enabled by technology innovations that reverses the previous industrial process logic". Industrial manufacturing equipment now "trains" products instead of just "processing" them [7]. As per experts, three potential areas of the industry's future are more clarity in energy system, demand flexibility, and energy efficiency. Industry 4.0 and the sustainable energy transition share basic qualities that might promote the transition. The sustainable development goal (SDG)'s energy, climate, and other aims might guide integrated approaches. Industry 4.0 prioritises energy efficiency.

2 Industry 4.0 Technologies

Industry 4.0 originally had nine pillars: big data, cyber-physical systems, the Internet of Things, 3D printing, robotics, simulation, augmented reality, cloud computing, and cyber security [8]. For better conceptual clarification, these technologies and their relevance have been explained in detail.

2.1 Big Data and Data Analytics

Data science analyses data sets using methods, scientific models, assumptions, and specialised equipment and software. Smart gadgets and advanced technologies like IoT, AI, social networking (SNS), and others have increased data sources, digital content, types, forms, and structure [9]. Five exabytes or five million terabytes of data are created daily [10]. Industry 4.0 generates data from sensors, log files, video/audio, network traffic, transactions, and social media [11].

Analysis complexity affects accuracy and quality. As firms receive more complicated data, they need technical, math/statistical, and business skills to design and provide an appropriate solution.

Thus, big data or heterogeneous data is generated daily and develops fast. Big data differs in volume, variety, honesty, velocity, and value [9, 12]. Big data gives industries and intelligent manufacturing several advantages, merits, and benefits via predictive and prospective insights. Therefore, to remain competitive, organisations should implement and apply contemporary advanced analytical tools, methods, methodologies, and applications for processing big data, obtaining insight, and recovering the value of each case's vital data. Big data analytics (BDA) uses parallel and analytic methods to analyse massive amounts of diverse, fast-changing data, making it easier to acquire, analyse, and manage vital information and statistics [9, 13]. The best method for companies to surpass competition, optimise operations, boost productivity, quality, and efficiency, and minimise operating costs is to use newly discovered information to deliver important insights and improve equipment servicing and maintainability [14]. Businesses must change their decision-making culture and remember that human understanding is still needed, even as big data and analytic technologies grow [12]. Figure 2 shows the data science model, Fig. 3 shows industry-wide applications and use cases, and Fig. 4 shows data analytics methodology, implementation, and measurement.

2.2 The Internet of Things (IoT)

Atzori et al. [16] describe IoT as "global network of uniquely addressable things employing standard communication protocols". Vermesan et al. [13] describe IoT as a self-configuring, global network architecture with open, standardised communication protocols. This network includes "things" with identities, physical traits, and virtual personalities [17]. Intelligent manufacturing uses the Industrial Internet of Things (IIoT). IIoT uses modern fundamental technologies to improve system efficiency [18]. IIoT services and apps increase industrial process and system planning, management, and scheduling [14, 19]. Networked devices will decentralise analytics and decision-making for real-time reactions [20]. Minimising unexpected downtime improves "availability, maintainability, operational efficiency, productivity, and new product time-to-market". It boosts industry [19]. Khan et al. [21]

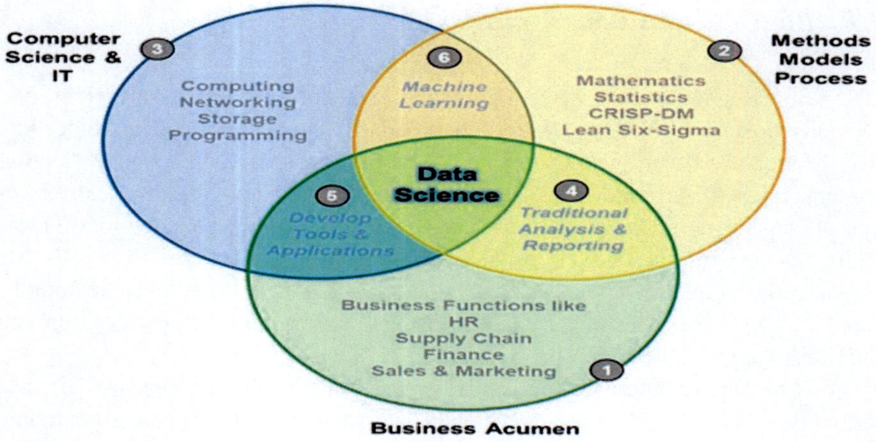

Fig. 2 The data science model [15]

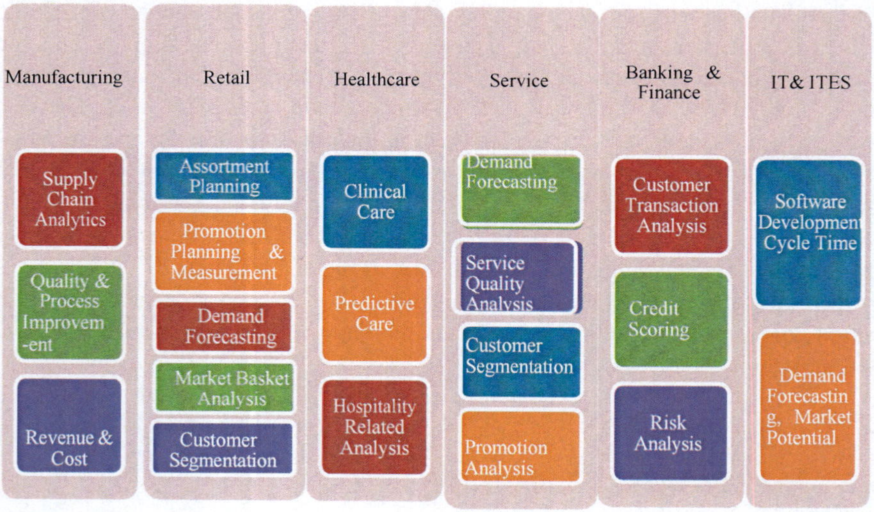

Fig. 3 Industry wide application and use cases [15]

examined "interoperability, standardisation, data and information secrecy, encryption, privacy, name and identity management, IoT greening, object and network security". Miorandi et al. examined communication, identification, distributed system, intelligence, security, data privacy, confidentiality, and trust [22]. Gubbi et al. studied "safe reprogrammable networks, privacy, QoS, energy-efficient sensing, architecture, and protocols, GIS-based visualisation, data mining, and cloud computing" [23]. Borgia includes "object mobility, M2M communications, device and data management, network architecture, system design, addressing, naming, traffic categorization, and security" [24]. As per Perera et al., privacy, data analytics, product and

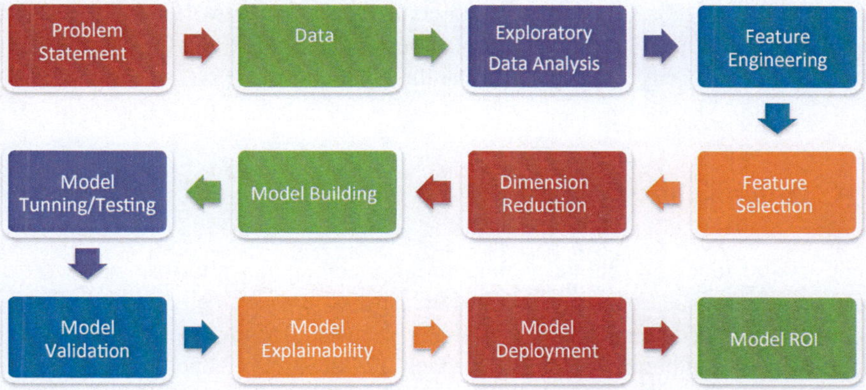

Fig. 4 A typical data analytics process, implementation, and measurement [15]

service interoperability, resource and energy management are key challenges [25]. AL-Faqaha et al. "examined IoT challenges and QoS demands such as availability, reliability, mobility, performance and management, scalability and interoperability, security, and privacy" [26]. Breivold and Sandstrom list IIoT problems as fault tolerance, functional safety, data latency and scalability, mixed-criticality, and secure real-time collaboration [27]. Lee considers data management, mining, security, and privacy the biggest IoT problems [28]. Sadeghi et al. [29] evaluated IIoT security, privacy, and attack weaknesses.

2.3 The Cloud

Benchmarking, colour management, and remote services utilise Cloud. Cloud computing, sometimes known as "Cloud", is a kind of outsourcing that uses many computer servers and resources to provide computer programmes, high-level services, and resources on demand or on a paper-cycle basis [30]. Wang et al. define cloud computing as "A set of network-enabled services that offer scalable, QoS-guaranteed, frequently customisable, and economical computing infrastructure on demand that can be accessed in a simple and broad way" [31]. Cloud computing services include SaaS, PaaS, and IaaS, which provide different levels of solution stack virtualisation and management [30]. Advanced apps and services that expand with users are one of the biggest benefits of cloud computing [32]. Consumers and companies may quickly access "Cloud" applications, programmes, and services with little administrative effort and from anywhere and at any time. Thus, industrial companies use cloud-based customer relationship management (CRM) and human resources management (HRM) apps to enhance critical processes. Cloud computing also eliminates the hassle of creating and maintaining an IT infrastructure and the upfront costs of the "Pay-as-you-go" approach [32]. Letting firms start as small and expand as demand rises [33, 34]. Due to cloud computing's fast growth, a wide range of applications and many advantages, businesses of all sorts are quickly adopting it to increase their capacity and capabilities at the lowest cost [34].

2.4 Autonomous Robots

Industry 4.0 relies on robotic manufacturing methods that prioritise safety, flexibility, diversity, and cooperation. More industrial robots are helping the Industrial Revolution with contemporary technologies. People and robots, work together using smart sensors and human–machine interfaces in Industry 4.0. Intelligent robots can be controlled by remote [35].

Industry 4.0 pioneers include some cutting-edge robotics. The lightweight Kuka LBR IIWA (Industrial Intelligent Work Assistant) is designed for human–robot cooperation on delicate tasks. It can independently check, optimise, and record its results while connected to the cloud [36]. Bosch also offers the APAS family robot system, which includes the APAS assistant, inspector, and base, for agile and flexible production based on a production system that can be easily retooled [37]. Rethink Robotics' Baxter makes interactive packages. BioRob Arm works near humans. Automated equipment performs repetitive operations quickly and accurately in locations where humans cannot reach easily [11].

2.5 Augmented Reality

Augmented Reality (AR) is used in computer and video games, social networking app filters, education, and learning to merge digital and real-world information. Milgram and Kishino [38] defined augmented reality as the interaction between real space, virtual space, and any intermediate mixed space in 1994. The AR Tool Kit was originally released outside US academic institutions at SIGGRAPH in 1999, launching the AR company. Two years later, it was released as open-source software. Smartphones and tablets have all the sensors and processing units needed to design and launch augmented reality apps. The Emacula contact lenses by Innoyega, the Vuzix blade 3000 AR glasses, and the Meta 2 AR headset are examples of emerging kinds of AR gadgets. Augmented reality's fast growth and widespread usage suggest a major social influence. AR in industry accelerates and improves product design and manufacturing development by boosting communication.

Some of the industrial AR applications: human–robot collaboration, maintenance-assembly-repair, training, product inspection, and building surveillance. Human–robot collaboration uses AR to create industrial robot interfaces. AR boosts maintenance-assembly-repair productivity. AR enhances skills in training. Inspectors may find product flaws using a sophisticated and adaptive AR system. Finally, augmented reality simplifies facility issues in building monitoring.

2.6 Cyber Security

Cyber-Physical Systems (CPS) underpin Industry 4.0 [39]. CPS automate physical reality operations using computer and communication infrastructures [40, 41]. A CPS are a network-capable embedded system. The "Internet of Things" includes internet-connected CPS. CPS connect devices, unlike standard embedded systems [41]. In today's digitally networked culture, CPS provide information and services everywhere.

CPS, the Internet of Things, and services have started the fourth Industrial Revolution. Germany leads CPS with almost 20 years of experience. Internet-enabled objects provide new services including cost-effective and efficient Internet-based diagnostics, maintenance, and operation. It also helps to adopt new business models, operational concepts, and intelligent controls while focussing on the user and their needs [42].

2.7 3D Printing

3D printing, known as additive manufacturing (AM) in Industry 4.0, can create complex metal or plastic structures [8]. AM, unlike subtractive manufacturing, combines materials to create a whole assembly using 3D model data created using particular software tools, frequently layer by layer [43, 44]. It allows varied production locations, reduced transportation, and inventories [11]. Due to technological breakthroughs, additive manufacturing in industry is growing despite worries about its mass production practicality. Since it can quickly produce precise, strengthened, and intricate items, it may replace traditional manufacturing procedures in the future. AM will improve technologies. Metal additive manufacturing is popular in this new age because metals are the most widely utilised industrial material [14, 16–19]. Digitalisation and the need to reduce product life cycles are attracting more sectors. Technologies like fused deposition, selective laser melting, and selective laser sintering are also covered [11].

2.8 Simulation

Product development and production are simulated using real-time data. Simulation can provide real-time data to speed up and improve testing. This optimises procedures and settings before production. It may speed up and improve product quality also. Plant operations need extensive simulation. Simulation may ensure product quality and reduce market price volatility [11]. It can decrease error-related downtime. Simulation helps Industry 4.0 to create exploratory and planning models for better decision-making, design, and operation of complex systems.

3 Opportunities, Issues, and Challenges in Industry 4.0

These new technologies enable real-time monitoring to improve product quality and encourage innovation across a wide range of applications, which has a stronger influence on economic growth. Automation and optimisation improve output, provide economic advantages through lower transaction and transit costs, predictive and remote maintenance, effective use of resources and staff, energy-efficient and ecologically friendly production system, and many more.

In the age of Industry 4.0, developing nations must keep up with the technology. Digital strategies demand technological knowledge and a business-friendly environment. Establishing places for discussion, information exchange and experience sharing would help society and government firms to boost up the development of emerging technologies for inclusive and sustainable industrial and economic progress.

These new technologies will have numerous benefits but also drawbacks. Researchers found many concerns and impediments, including: many present systems lack autonomy, most network protocols lack capacity, and many sectors have yet to secure data quality and integrity [45]. There is no standard technique for data entity annotations [46], complicated system modelling and analysis are not yet viable [45], and changing production routes to accept a big dynamical reconfiguration for individualised and customised goods is challenging [11], CPS stability, new talent development, privacy, ethics, and management change, outdated international laws, and data quality.

4 Conclusion and Future Work

The fourth Industrial Revolution of the twenty-firstst century enables intelligent, effective, personalised and tailored production. This study examines Industry 4.0, its development, and its components. Industrial progress has been extensively documented. The benefits, drawbacks, and challenges of Industry 4.0 are discussed with its nine pillars. Since Industry 4.0 is new, future challenges will undoubtedly rise. Industry 4.0 may change an organisation's value chain and provide it with a competitive edge in the global market. It promotes social growth, good governance, and transparency.

References

1. Weyer S, Schmitt M, Ohmer M, Gorecky D (2015) IFAC-Papers on Line 48(3):579–584
2. Lasi H, Fettke P, Kemper HG, Feld T, Hoffmann M (2014) Industry 4.0. Bus Inf Syst Eng 6:239–242

3. Chen B, Wan J, Shu L, Li P, Mukherjee M, Yin B (2017) Smart factory of industry 4.0: key technologies, application case, and challenges. IEEE Access 6:6506–6519
4. Acemogu D (2002) Technical change, inequality, and the labor market. J Econ Lit 40(1):7–72
5. Tunzelmann NV (2003) Historical coevolution of governance and technology in the industrial revolutions. Struct Chang Econ Dyn 14(4):365–384
6. Schuh G et al (2013) Sust. Increase of overhead productivity due to cyber-physical system. In: Seliger G (ed) Proceedings of the 11th global conference on sustainable manufacturing
7. Kumar A, Kumar S (2020) Industry 4.0: evolution, opportunities and challenges. Int J Res Bus Stud 5(1). ISSN: 2455–2992
8. Yang F, and Gu S (2021) Industry 4.0, a revolution that requires technology and national strategies. Complex & Intell Syst 7:13, 11–1325
9. Gahi Y, Guennoun M, Mouftah HT (2016) Big data analytics: security and privacy challenges. In: 2016 IEEE symposium on computers and communication (ISCC). IEEE, pp 952–957
10. Proctor M, Wilkins J (2019) 4.0 sight—digital industry around the world. Technical report, EU Automation
11. Vaidya S, Ambad P, Bhosle S (2018) Industry 4.0—a glimpse. Procedia Manuf 20:233–238
12. McAfee A, Brynjolfsson E, Davenport TH, Patil DJ, Barton D (2012) Big data: the management revolution. Harv Bus Rev 90(10):60–68
13. Parwez MS, Rawat DB, Garuba M (2017) Big data analytics for user-activity analysis and useranomaly detection in mobile wireless network. IEEE Trans Industr Inf 13(4):2058–2065
14. Gilchrist A (2016) Industry 4.0: the industrial internet of things. Apress, New York, NY. ISBN: 1484220463
15. Kaliraj P, Devi T (eds) (2022) Big data applications in industry 4.0, 1st edn. Auerbach publications
16. Atzori L, Iera A, Morabito G (2010) The internet of things: a survey. Comput Netw 54(15):2787–2805
17. Vermesan O, Friess P, Guillemin P, Gusmeroli S, Sundmaeker H, Bassi A, Doody P (2011) Internet of things strategic research roadmap. In: Internet of things-global technological and societal trends, vol 1. pp 9−52
18. Lampropoulos G, Siakas K, Anastasiadis T (2018) Internet of things (IoT) in industry: contemporary application domains, innovative technologies and intelligent manufacturing. Int J Adv Sci Res Eng 4(10):109–118
19. Bi Z, Da Xu L, Wang C (2014) Internet of things for enterprise systems of modern manufacturing. IEEE Trans Industr Inf 10(2):1537–1546
20. Şen KO, Durakbasa MN, Baysal MV, Şen G, Baş G (2018) Smart factories: a review of situation, and recommendations to accelerate the evolution process. In: The international symposium for production research. pp 464–479
21. Khan R, Khan SU, Zaheer R, Khan S (2012) Future internet: the internet of things architecture, possible applications and key challenges. In: 2012 10th international conference on frontiers of information technology. IEEE, pp. 257–260
22. Miorandi D, Sicari S, De Pellegrini F, Chlamtac I (2012) Internet of things: vision, applications and research challenges. Ad Hoc Netw 10(7):1497–1516
23. Gubbi J, Buyya R, Marusic S, Palaniswami M (2013) Internet of things (IoT): a vision, architectural elements, and future directions. Futur Gener Comput Syst 29(7):1645–1660
24. Borgia E (2014) The internet of things vision: key features, applications and open issues. Comput Commun 54:1–31
25. Perera C, Liu CH, Jayawardena S, Chen M (2014) A survey on internet of things from industrial market perspective. IEEE Access 2:1660–1679
26. Al-Fuqaha A, Guizani M, Mohammadi M, Aledhari M, Ayyash M (2015) Internet of things: a survey on enabling technologies, protocols, and applications. IEEE communications surveys & tutorials 17(4):2347–2376
27. Breivold HP, Sandström K (2015) Internet of things for industrial automation--challenges and technical solutions. In: 2015 IEEE international conference on data science and data intensive systems. IEEE, pp 532–539

28. Lee J, Bagheri B, Kao HA (2015) A cyber-physical systems architecture for industry 4.0-based manufacturing systems. Manuf lett 3:18–23
29. Sadeghi AR, Wachsmann C, Waidner M (2015) Security and privacy challenges in industrial internet of things. In: 2015 52nd ACM/EDAC/IEEE design automation conference (DAC). IEEE, pp 1–6
30. Bhardwaj S, Jain L, Jain S (2010) Cloud computing: a study of infrastructure as a service (IAAS). Int J Eng Inf Technol 2(1):60–63
31. Wang L, Von Laszewski G, Younge A, He X, Kunze M, Tao J, Fu C (2010) Cloud computing: a perspective study. N Gener Comput 28(2):137–146
32. Lu Y, Cecil J (2016) An internet of things (IoT)-based collaborative framework for advanced manufacturing. Int J Adv Manuf Technol 84(5–8):1141–1152
33. Zhang Q, Cheng L, Boutaba R (2010) Cloud computing: state-of-the-art and research challenges. J Internet Serv Appl 1(1):7–18
34. Zhong RY, Xu X, Klotz E, Newman ST (2017) Intelligent manufacturing in the context of industry 4.0: a review. Eng 3(5):616–630
35. Roland Berger Strategy Consultants (2014) Industry 4.0, The new industrial revolution: how Europe will succeed. In: International conference the next industrial revolution manufacturing and society in the XXI century, Turin, November 14–15.
36. Bauer A (2016) KUKA LBR Iiwa. Kuka Roboter GmbH, Augsburg
37. Robert Bosch GmbH (2015) Changeable automation with the APAS family. Stuttgart
38. Milgram P, Kishino F (1994) A taxonomy of mixed reality visual displays. IEICE Trans Inf Syst 77(12):1321–1329
39. Ji C, Shao Q, Sun J, Liu S, Pan L, Wu L, Yang C (2016) Device data ingestion for industrial big data platforms with a case study. Sensors 16(3):279
40. Broy M (2010) Cyber Physical Systems-Wissenschaftliche Herausforderungen bei der Entwicklung. Verlag: Springer Berlin Heidelberg; Auflage: 1st Edition
41. Lee E (2008) Cyber physical systems: design challenges. University of California, Berkeley Technical Report No. UCB/EECS-2008–8. http://www.eecs.berkeley.edu/Pubs/TechRpts/2008/EECS-2008-8.html
42. Bohler TM (2012) Industrie 4.0–Smarte Produkte und Fabriken revolutionieren die Industrie. In: Produktion Magazin, 10. Mai 2012; Abgerufen am 5. September 2012
43. Mehrpouya M, Dehghanghadikolaei A, Fotovvati B, Vosooghnia A, Emamian SS, Gisario A (2019) The potential of additive manufacturing in the smart factory industrial 4.0: a review. Appl Sci 9(18):3865
44. Prashar G, Vasudev H (2021) A comprehensive review on sustainable cold spray additive manufacturing: state of the art, challenges and future challenges. J Clean Prod 310:127606
45. Wang S, Wan J, Li D, Zhang C (2016) Implementing smart factory of industrie 4.0: an outlook. Int J Distrib Sens Netw 1–1
46. Thoben KD, Wiesner S, Wuest T (2017) Industrie 4.0 and smart manufacturing—a review of research issues and application examples. Int J Autom Technol 11(1):4–16

AI Control of EMG Sensor Data for 3D Printed Prosthesis Hand

Gunasundar Paddam, Vishal Francis, and Narendra Kumar

1 Introduction

Around 40 million amputees live in underdeveloped nations, according to the World Health Organization. Additionally, prostheses are necessary for 0.5% of the world's population [1, 2]. Amputation significantly limits functionality, negatively impacts social relationships, and undermines a person's confidence. A prosthetic for the upper limb is highly expensive, and they are hard to come by in many remote places. Many amputees opt not to utilize prosthetic hands because of these circumstances. In addition, many of the existing prostheses lack aesthetic appeal and are too heavy for an amputee. Children from underdeveloped nations who eventually outgrow their prostheses have it even worse.

Another key element of a successful prosthetic hand is customization, therefore, an upper extremity must be amputated in order to create the prosthesis. The fabrication of economical and dependable prostheses can be a feasible remedy for rural communities in this situation.

The fabrication of any complex and customized shape geometry using 3D printing is facilitated and is a prerequisite for the creation of prosthetics. The major features of 3D printing that makes it a useful tool for prosthesis development are speed, customization, growth of amputees, and comfort. Prosthetics created using 3D printing can be created in a matter of days as opposed to the longer time frame which is the case with other traditional methods. To meet the patient's unique needs, customized prostheses can be created using the 3D printing method without the use

G. Paddam · V. Francis (✉)
School of Mechanical Engineering, Lovely Professional University, Jalandhar, India
e-mail: vishal.24813@lpu.co.in

N. Kumar
Department of Industrial and Production Engineering, Dr B R Ambedkar Natrional Institute of Technology, Jalandhar 144027, India

V. S. Sharma et al. (eds.), *Machining and Additive Manufacturing*, Lecture Notes in Mechanical Engineering, https://doi.org/10.1007/978-981-99-6094-1_8

of tools. Since prostheses for growing children must be replaced after a few years, they should be cost-effective [3–5].

Many families may find affordable 3D printed prosthetics to be a good choice. Additionally, a custom-made socket can be created based on the structure of the amputee's hand to increase comfort.

Depending on how they operate, prosthetic hands can be divided into three categories. The first has either no or very little utility, and it is passive. Another sort of prosthetic hand that doesn't require electronics and can have a limited range of motion is body powered. Another class of prosthetic hand that offers more degrees of flexibility than other prosthetic hand categories is external powered.

The use of various printing materials, such as metal powders, thermoplastic polymers, photo polymers, molding sands for casting, etc., is made possible by 3D printing technology. The 3D printing process is distinct from traditional production methods. The material is removed via conventional machining. The reverse is the fundamental idea behind additive manufacturing. By adding layers, the final 3D shape "grows." The term "additive manufacturing" was used to describe this type of production (AM).

For a diverse range of reasons, 3D printing has been successful in producing prostheses. The time, cost, and weight required to manufacture prosthetics are significantly reduced when they are created using 3D printing. The cost of 3D printed prosthetics is low, the production time is quick, and the materials are easily accessible. 3D printing is classified into seven broad categories by the International Standards Organization as illustrated in Fig. 1.

The control of the prosthetic hand is very crucial for improved functionality of the printed hand. Over the past decades, the advancements in the control strategies and the use of sensors for feedback have improved the functionality of prosthetic hands.

The incorporation of electromyographic (EMG) sensors for the control of the printed hand has further opened the scope for better control and functionality.

However, the signals captured from the muscle's sensor need to be classified for controlling the actuators. Therefore, the pattern needs to be recognized and features extracted to make full use of EMG sensors for 3D printed prosthetic control.

Various Artificial Intelligence/Machine Learning (AI/ML) algorithms are used for this purpose. The present paper discusses the various methods used for the control of prosthetic hands along with the data acquisition from muscle sensors and their placement. The 3D printed prosthetic hand with improved functionality and control can open avenues for developing low-cost functional prosthesesand.

2 EMG Sensor

An electromyography sensor, also known as an EMG sensor, detects small electrical signals produced by your muscles as you move them. It works on the principle of electrodes are placed close to muscle groups to record EMG signals. The length of the muscle shortens as it is activated, and the muscle, skin, and electrodes move

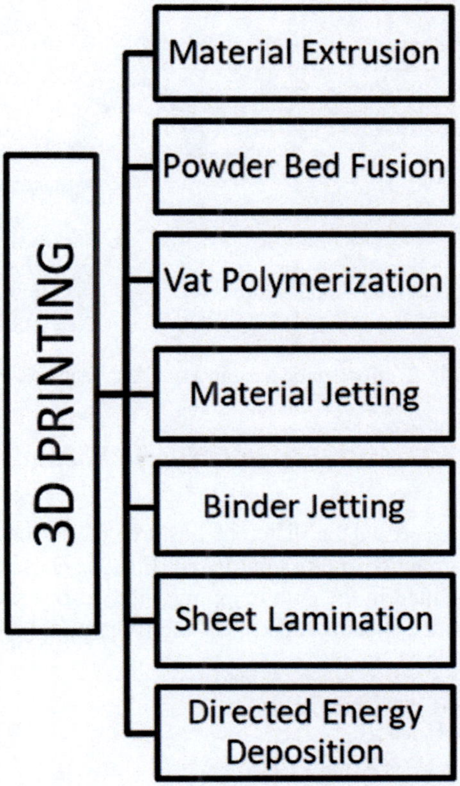

Fig. 1 Classification of 3D printing process

in relation to one another. The electrodes will show some movement artifacts at that time [6]. EMG results can reveal nerve dysfunction, muscle dysfunction, or problems with nerve-to-muscle signal transmission. By measuring the response of muscles to stimulation, the EMG is used to assess muscle health. When a patient exhibits unexplained muscle weakness, doctors can use this information to rule out multiple sclerosis and other diseases [7].

2.1 Myoelectric Hand

Myoelectric prosthetic hand is one that is controlled by myoelectric sensors. It has a number of sEMG sensors to operate various prostheses portions. Figure 2 illustrates a block diagram depicting the operation of the myoelectric hands. The shaded area represents amputation, which is replaced by a prosthetic hand [8].

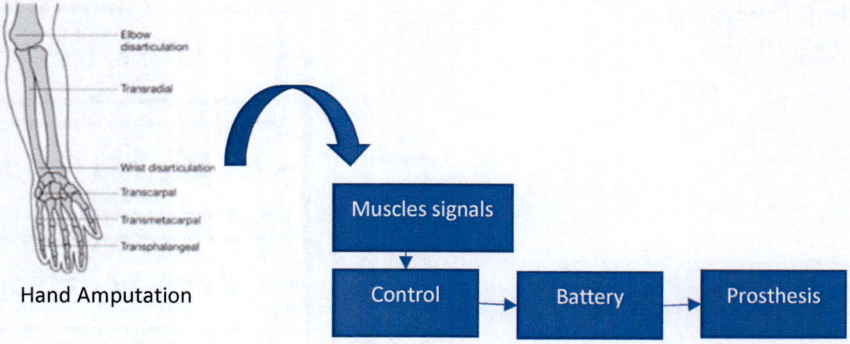

Fig. 2 Myoelectric hand function representation [8]

2.2 *Postures and Muscle Locations*

Based on the different postures of the hand, various muscles of the forearm are stretched and/or contracted (Fig. 3). These different hand postures have been represented in the data mapping with various muscles, which actuate the fingers [9, 10] since specific muscles of the forearm are required to move different fingers.

Fig. 3 Finger and muscle location [10]

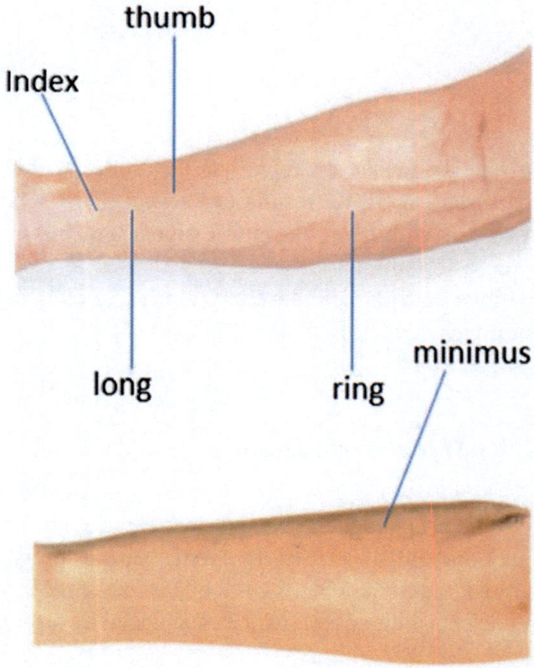

2.3 EMG in Prostheses

Alexander Calando et al. conducted a review of commercially available anthropomorphic myoelectric prosthetic hand and concluded that pattern-recognition-based microcontrollers for prostheses are still uncommon due to funding issues in this area [11]. Chenguang Yang et al investigated the human–robot interaction system and concluded that EMG-based results yielded improved performance [12]. In this way, Lingling Chen et al explored the selection of EMG sensors based on motion analysis and identified the directed network and key muscle groups. Using convergent cross-mapping, an absolute contrast between strong restrained movement and unrestrained movement exhibited [13].

On the other hand, Seulah Lee et al. developed a knit band sensor for myoelectric control, whereas most sEMG sensors use disposable electrodes and have an accuracy of 87.9%. The developed knit band has a higher signal to noise ratio (17.91 dB) than disposable electrodes (13.3 dB) and can classify motion with 93.2% accuracy [9].

3 Artificial Intelligence

Artificial intelligence (AI) is the competence of a computer or robot that is self-controlled to perform functions that are typically carried out by humans. The rapid development of AI has gained popularity. Almost all industries are using AI to forecast various factors from "Big Data" sets [14]. Here the study will be conducted using the purposive sampling approach, with analysis based on the resources relevant to the subject chosen. The element here is to assess the relevance and rising needs of AI and ML. A systematic comprehension of the problem necessitates engaging some study participants to assess the level of performance and resource management [15]. Figure 4 illustrates the classification of artificial intelligence.

3.1 Machine Learning

Machine learning incorporates aspects of mathematics, statistics, and computer science. It is the exploration of computer algorithms. There are two forms of machine learning. Learning methods: supervised and unsupervised. Machine learning is a subtechnique of artificial intelligence as shown in Fig. 4. According to many authors, machine learning is a field that enables systems to learn on their own. Here, training data sets will be used to train the machine, and test data will be used to evaluate it later. There are still several areas where this sector has to expand in which the invention is required [16–18].

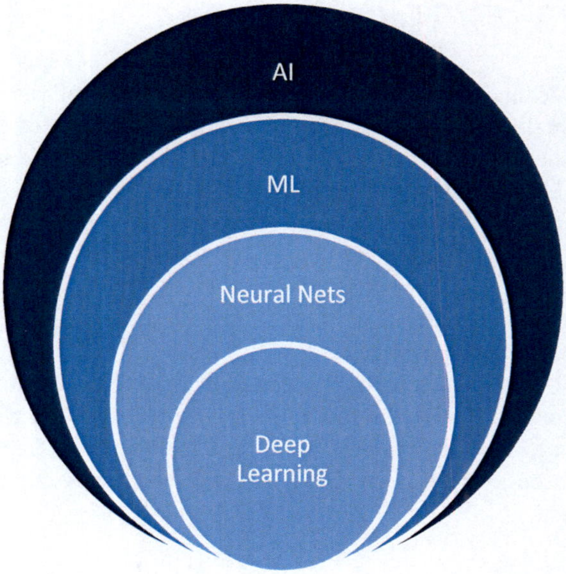

Fig. 4 Artificial intelligence classification

3.2 *Supervised Machine Learning*

Supervised learning is a technique in which the model is trained using labeled data, and the function mapping is based on the known output utilized during training. Algorithms then predict and classify data based on the specified labels. Its performance is determined by the number of correctly predicted predetermined labels. This category includes regression and classification [19, 20].

3.3 *Unsupervised Machine Learning*

Unsupervised learning is a technique in which the model is fed with unlabeled data. The objective in this case is to extract the structure and pattern from the input data. This category includes clustering and association [19, 20]. In this field of research, this division of technique RNN (recurrent neural network) has been implied. As a signal classification technique, the recurrent neural network is an unsupervised learning system that has the ability to remember input data due to internal memory. As a result, this approach is ideal for sequential data [21]. Figure 5 illustrates the sequence of capturing muscles and final control of prosthetic hand using ML models.

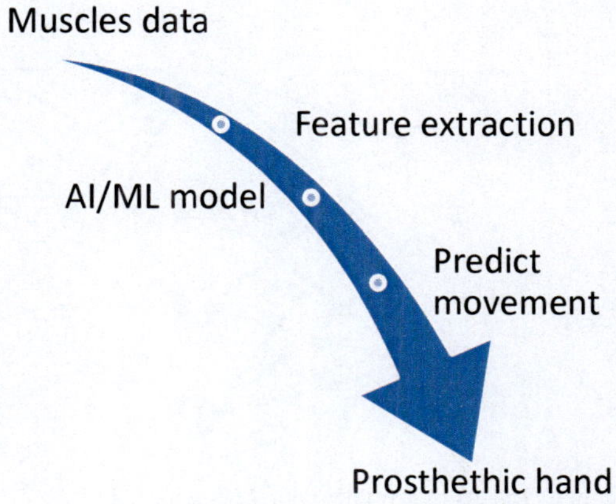

Fig. 5 Signal classifier in the prosthetic hand [21]

3.4 *Neural Networks*

A neural network is a nonlinear mathematical function that precisely turns a set of input variables into a set of output variables. This process is guided by a set of parameters known as weights. The needed mapping can determine weights [18, 22]. In this area of investigation, this method is used.

3.5 *Deep Learning*

Deep learning is a subtype of machine learning in which neural networks learn from massive sets of data and have hundreds of representation layers. This method is employed in the recognition of objects, speech, and objects [23]. It employs artificial neural networks (ANN) representation learning, which is inspired by the human neural network. Figure 6 shows the hidden layers in ANN. It falls under both supervised and unsupervised learning [24]. The ANN approach has implications in the prosthetic hand field. Figure 7 illustrates the use of ANN as a signal classifier for prosthetic hands.

4 AI in Prosthetics

Artificial intelligence is being employed everywhere in the evolving globe. The large amount of information that we produce will assist us in developing ideal solutions by analyzing the relationships and patterns. Basmajian et al. stated that, because of rapid improvements in robotics technology, robots are expected to physically execute human interactions to support daily operations [26]. The data from several

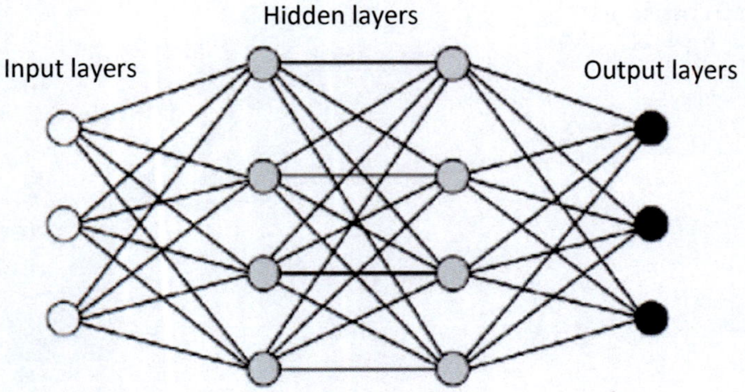

Fig. 6 Layers of artificial neural network [24]

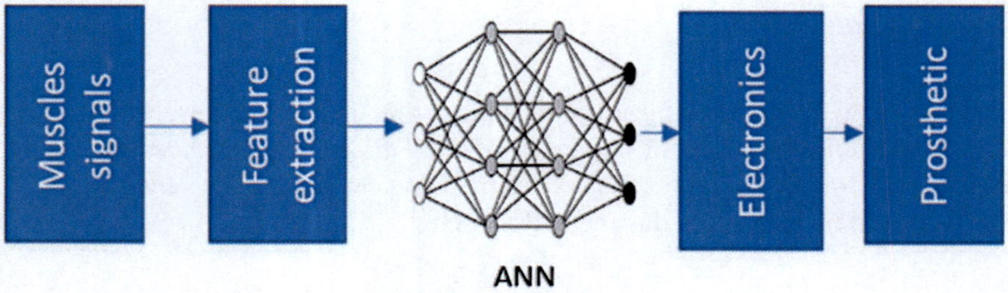

Fig. 7 Basic structure of prosthetic control using artificial neural network [25]

writers who implemented artificial intelligence in the field of prosthetics is as follows. Authors Diu Khu et al. have developed a real-time intuitive control of prostheses. By nerve data here, an AI agent uses a peripheral nerve interface to communicate the intent of an amputee. The recurrent neural network (RNN) was used in this study [27]. Another researcher Artal-Sevil et al. used a neural networks approach to map muscle data for a gesture detection system and achieved satisfactory results [28].

Bittibissi et al. implemented a recurrent neural network (RNN) based on long-term short-term memory (LTSM) and gated recurrent unit (GRU). According to the study's findings, GRU is the best option due to its simpler and faster model than the convolution LTSM model [29]. Another researcher, Rezwanul et al., created an ANN-based signal recognition model. Backpropagation is used in this network, which is based on the Levenberg Marquardt equation and achieved an 88.4% success rate [22]. Khomami et al. have investigated language recognition using IMU and EMG sensors on Persian signs. Surface electromyography (sEMG) and Inertial Measurement Unit (IMU) sensors were employed in the low-cost device that researchers designed and fabricated. Sign capture accuracy increased when these two sensors were used together and achieved 96.13 average accuracy, using the KNN classifier method [30]. KNN is a non-parametric, simple, and efficient supervised machine learning classification approach [31].

5 Conclusions

The 3D printed prosthetic hand with improved functionality and control is of utmost importance. Generally, the EMG sensors are utilized to collect data from muscles and map signals. This data needs to be processed for further control of the prosthetic hand which can improve the control and functionality. However, there is a need to increase the accuracy in control which is a very important aspect in prosthetics.

The inclusion of AI/ML techniques in control of the prosthetic hand can improve its functionality. AI approaches like KNN, RNN, and ANN are been employed, where ANN being the only supervised learning methodology. The accuracy can be significantly improved by employing these techniques. Further exhaustive studies are needed to fully explore and utilize the benefit of AI/ML techniques in control of prosthetic hands. It can be concluded that the use of AI/ML techniques for control of 3D printed prosthetic hands can be a viable solution for economical prostheses with adequate functionality.

References

1. Dally C, Johnson D, Canon M, Ritter S, Mehta K (2015) Characteristics of a 3D-printed prosthetic hand for use in developing countries. In: 2015 IEEE global humanitarian technology conference (GHTC). IEEE, pp 66–70
2. Baril M, Gosselin C, de Jesus Lima E, Arabiam A (2017) Development of a 3d printed prosthetic myoelectric hand driven by Dc actuators, vol 4. Technology Department State University of Feira de Santana, BA, Brazil
3. Dombroski CE, Balsdon MER, Froats A (2014) The use of a low cost 3D scanning and printing tool in the manufacture of custom-made foot orthoses: a preliminary study. BMC Res Notes 7(1):1–4
4. Muilenburg AL, LeBlanc MA (1989) Body-powered upper-limb components. In: Comprehensive management of the upper-limb amputee. pp 28–38
5. Zuniga JM (2018) 3D printed antibacterial prostheses. Appl Sci 8(9):1651
6. Chowdhury RH, Reaz MBI, Ali MABM, Bakar AAA, Chellappan K, Chang TG (2013) Surface electromyography signal processing and classification techniques. Sensors 13(9):12431–12466
7. Doheny EP, Goulding C, Flood MW, Mcmanus L, Lowery MM (2019) Feature-based evaluation of a wearable surface EMG sensor against laboratory standard EMG during force-varying and fatiguing contractions. IEEE Sens J 20(5):2757–2765
8. Hudgins B, Parker P, Scott RN (1993) A new strategy for multifunction myoelectric control. IEEE Trans Biomed Eng 40(1):82–94
9. Lee S, Kim M-O, Kang T, Park J, Choi Y (2018) Knit band sensor for myoelectric control of surface EMG-based prosthetic hand. IEEE Sens J 18(20):8578–8586
10. Unanyan NN, Belov AA (2019) Signal-based approach to EMG-sensor fault detection in upper limb prosthetics. In: 2019 20th international Carpathian control conference (ICCC). IEEE, pp 1–6
11. Calado A, Soares F, Matos D (2019) A review on commercially available anthropomorphic myoelectric prosthetic hands, pattern-recognition-based microcontrollers and sEMG sensors used for prosthetic control. In: 2019 IEEE international conference on autonomous robot systems and competitions (ICARSC). IEEE, pp. 1–6

12. Yang C, Zeng C, Liang P, Li Z, Li R, Chun-Yi S (2017) Interface design of a physical human–robot interaction system for human impedance adaptive skill transfer. IEEE Trans Autom Sci Eng 15(1):329–340
13. Chen L, Liu X, Xuan B, Zhang J, Liu Z, Zhang Y (2021) Selection of EMG sensors based on motion coordinated analysis. Sensors 21(4):1147
14. Xu Y, Zhang X, Li H, Zheng H, Zhang J, Olsen MS, Varshney RK, Prasanna BM, Qian Q (2022) Smart breeding driven by big data, artificial intelligence and integrated genomic-enviromic prediction. Mol Plant
15. Gong Y, Chen K, Niu T, Liu Y (2022) Grid-based coverage path planning with NFZ avoidance for UAV using parallel self-adaptive ant colony optimization algorithm in cloud IoT. J Cloud Comput 11(1):29
16. Hurwitz J, Kirsch D (2018) Machine learning IBM limited edition. Accessed 22 Mar 2021
17. Jordan MI, Mitchell TM (2015) Machine learning: trends, perspectives, and prospects. Science 349(6245):255–260
18. Bishop CM (1994) Neural networks and their applications. Rev Sci Instrum 65(6):1803–1832
19. Alloghani M, Al-Jumeily D, Mustafina J, Hussain A, Aljaaf AJ (2020) A systematic review on supervised and unsupervised machine learning algorithms for data science. Supervised Unsupervised Learn Data Sci, 3–21.
20. Ozgur A (2004) Supervised and unsupervised machine learning techniques for text document categorization. Unpublished Master's Thesis, İstanbul: Boğaziçi University
21. Yin W, Kann K, Yu M, Schütze H (2017) Comparative study of CNN and RNN for natural language processing. arXiv:1702.01923
22. Ahsan MR, Ibrahimy MI, Khalifa OO (2011) Electromygraphy (EMG) signal based hand gesture recognition using artificial neural network (ANN). In: 2011 4th international conference on mechatronics (ICOM). IEEE, pp 1–6
23. LeCun Y, Bengio Y, Hinton G (2015) Deep learning. nature, 521 (7553), 436–444. Google Sch Google Sch Cross Ref Cross Ref 25
24. Spiers DL (2016) Facial emotion detection using deep learning
25. Englehart K, Hudgins B, Parker PA (2001) Multifunction control of powered prostheses using the myoelectric signal. In: Intelligent technologies for rehabilitation. CRC Press, Boca Raton, FL1–61
26. Basmajian JV (1985) Their function revealed by electromyography. Muscle Alive 212
27. Luu DK, Nguyen AT, Jiang M, Drealan MW, Xu J, Wu T, Tam W et al (2022) Artificial intelligence enables real-time and intuitive control of prostheses via nerve interface. IEEE Trans Biomed Eng 69(10):3051–3063
28. Artal-Sevil JS, Perez-Cebolla FJ, Pascual-Acon A, Dominguez JA (2020) Control of a bionic hand based on neural networks and improved gesture recognition techniques using multiple EMG sensors. In: 2020 XIV technologies applied to electronics teaching conference (TAEE). IEEE, pp 1–8
29. Bittibssi TM, Genedy MA, Maged SA (2021) sEMG pattern recognition based on recurrent neural network. Biomed Signal Process Control 70:103048
30. Khomami SA, Shamekhi S (2021) Persian sign language recognition using IMU and surface EMG sensors. Measurement 168:108471
31. Guo G, Wang H, Bell D, Bi Y, Greer K (2003) KNN model-based approach in classification. In: On the move to meaningful internet systems 2003: CoopIS, DOA, and ODBASE: OTM confederated international conferences, CoopIS, DOA, and ODBASE 2003, Catania, Sicily, Italy, November 3–7, 2003. Proceedings. Springer Berlin Heidelberg, pp 986–996

Methodology of Rapid Prototyping and Moulding of Net-Shaped Micro-structures Developed Using Lifting Plate Hele-Shaw Flow on Conical Surfaces

Sanmit Pandey, Yogini Chaudhari, Pranjal Pawar, and Kiran Suresh Bhole

1 Introduction

Fractals are one of those discoveries, which humans made by closely observing nature. Structures like snowflakes, human veins and veins of a leaf [1] have thrilled us to work on fractals. Humans have there on made multiple attempts to mimic fractals in a laboratory; many of which yielded fruitful results. These fractals can be used for benefits such as cache locality [2], micro-mixing [3], bio-mimicking [4] and many more. When we use the lifting plate Hele-Shaw flow methodology, fractals are formed as a result of instabilities occurring due to the interaction of high-viscosity and low-viscosity fluids. In the context of manufacturing net-shaped micro-structures, this method proves to be far superior to bulk lithography[5], microstereolithography[6], interference lithography and so on [7]. Extending these benefits from 2-D flat plates to 3-D curved conical plates shall have various benefits. The fractal formation process can also be controlled using certain methods, which further aids the use case of fractals [8]. In this work, we shall be doing these experiments using conical plates in order to get three-dimensional fractals [9].

S. Pandey (✉) · Y. Chaudhari (✉) · P. Pawar (✉) · K. S. Bhole
Department of Mechanical Engineering, Sardar Patel College of Engineering, Andheri, Mumbai 400078, India
e-mail: m1910046@spce.ac.in

Y. Chaudhari
e-mail: m1910010@spce.ac.in

P. Pawar
e-mail: m1910051@spce.ac.in

V. S. Sharma et al. (eds.), *Machining and Additive Manufacturing*, Lecture Notes in Mechanical Engineering, https://doi.org/10.1007/978-981-99-6094-1_9

2 Experimentation

2.1 Fractal Formation Methodology

The fractals are formed using the setup shown in Fig. 1. In a study, the non-Newtonian fluid, which in our case happens to be toothpaste[10] is filled between the narrow gap of two closely fitting conical surfaces (refer Fig. 2). The upper plate is then brought down till its distance from the lower plate is very small as close as 0.1 mm after which the upper plate is slowly [11] moved back to its original position. This is called the Lifting Plate Hele-Shaw flow [12]. The velocity is varied using a stepper motor[13] which is controlled by a GUI made in the pronterface (refer Fig. 1). The small gap position is responsible for promoting Saffman-Taylor [14] instabilities [15], which corresponds to fractal formation [16].

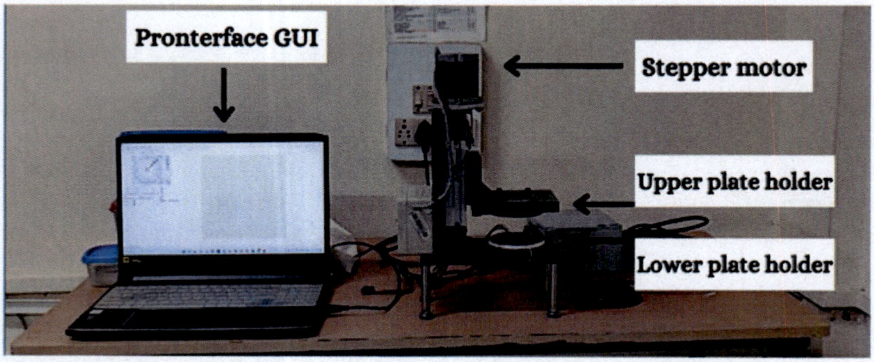

Fig. 1 Fractal formation setup

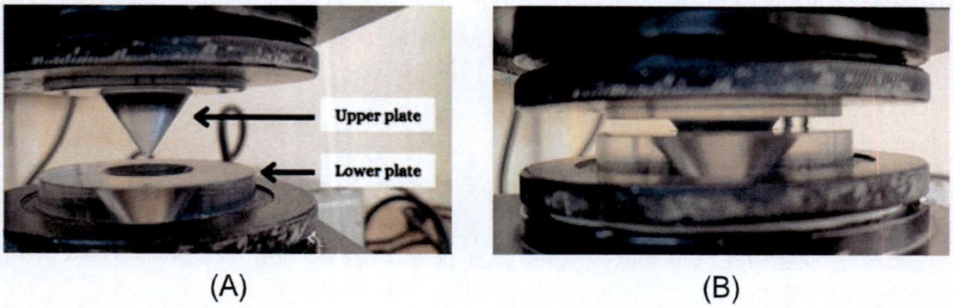

Fig. 2 Position of plates **a** initial position **b** small gap position

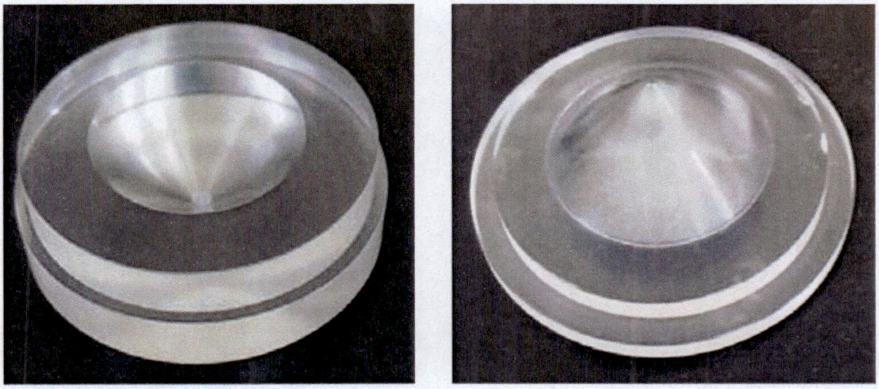

(A) (B)

Fig. 3 Conical plates **a** lower plate **b** upper plate

2.2 *Conical Plates*

Considering the effects of surface roughness and the benefits obtained through the designed setup in our investigation using a transparent material, the study has chosen to use acrylic plates (refer Fig. 3a and b). The dimensions of these plates are constrained largely to the setup. For the work, we have fabricated various types of conical plates, with the solid angle being the only intended variable parameter.

2.3 *Moulding Methodology*

Moulding of these fractals is crucial to make any substantial use of our fractals either by using a mechanically stronger alternative or for uses like micro-mixing [3]. With perspectives of mould removal, economic feasibility and ease of handling, we have chosen to use Polydimethylsiloxane (PDMS) as our mould material. It has been widely used in microfluidics. We mix the PDMS polymer along with PDMS curing agent and then the mixture undergoes degassing for the removal of air bubbles after which it can be used (refer Fig. 4).

Fig. 4 Degassing of PDMS mixture

3 Results and Discussions

3.1 *Fractal Formation*

On experimenting with all conical plates, fractal formation is evident. These fractals are not completely symmetric as seen in 2-D flat plates[17] but show a close resemblance to each other (refer Fig. 5). The fractals were very similar to the ones attained on flat plates, but they are imprinted along the curvature of the cone.

With experiments conducted using other non-Newtonian fluids, it was also established that higher viscosities tend to give better fractals. The thickness of these fractals had a lot to do with the amount of non-Newtonian fluid used and was also a crucial parameter for fractal formation.

3.2 *Mould Formation*

Moulding around a conical plate is viable. The fractals are suitably imprinted on the surface. Unlike flat plates, moulds made by the upper plate and lower plate are completely different (refer Fig. 6). The mould made on the upper plate resembles the fractals obtained on the lower plate and vice versa.

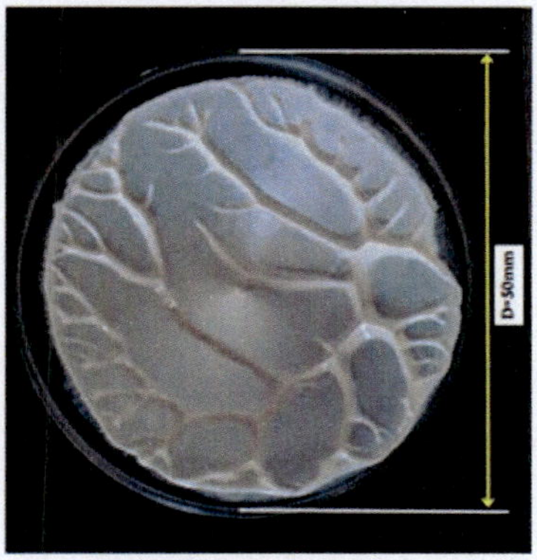

(A) Top view of the lower plate

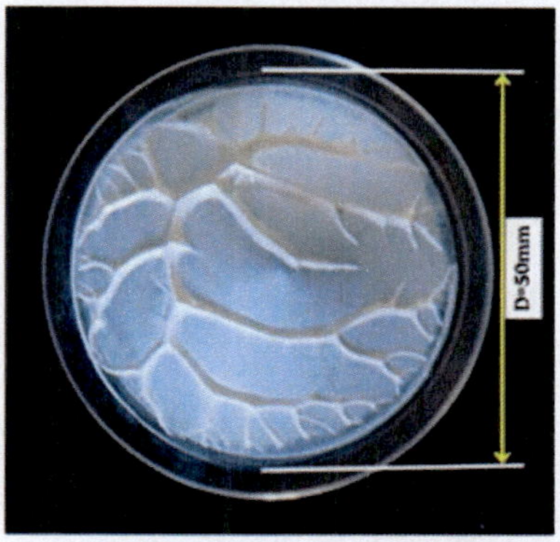

(B) Top view of the upper plate

(C) Curvature of fractals on the lower plate

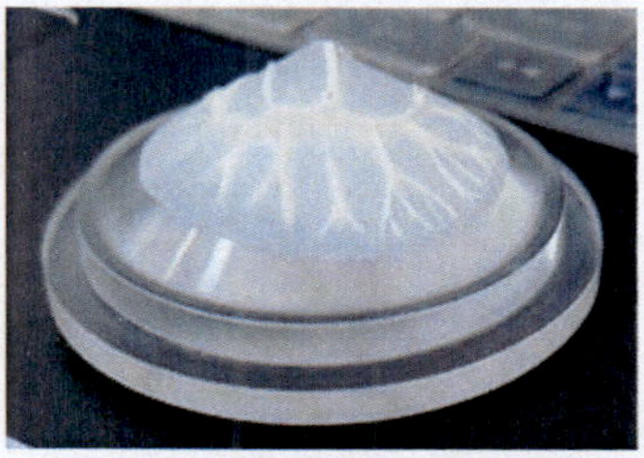

(D) Curvature of fractals on the upper plate

Fig. 5 Results of fractals formation

Fig. 6 Moulds formed of fractals using PDMS

4 Conclusions

The lifting plate Hele-Shaw methodology was used to obtain fractals on curved conical surfaces with varying solid angles, which were then processed to make moulds of these fractals for various applications. There were observations drawn and demonstrated regarding the differences between the fractals formed over a 2-D Flat plate and a 3-D conical plate with discussions around the curvature of the fractals formed.

The moulding techniques used were also discussed and demonstrated for conical fractals obtained using lifting plate Hele-shaw flow.

Acknowledgements The authors acknowledge support for this work by the Science and Engineering Research Board (SERB), Government of India through Project Grant CRG/2021/000747.

References

1. Kale BS et al (2022) Effect of polygonal surfaces on development of viscous fingering in lifting plate Hele-Shaw cell. Int J Interact Des Manuf IJIDeM. https://doi.org/10.1007/s12008-022-01030-9
2. Bilardi G, D'Alberto P, Nicolau A (2001) Fractal matrix multiplication: a case study on portability of cache performance. In: algorithm engineering, Berlin, Heidelberg. pp 26–38
3. Kale BS, Bhole KS, Bhole D, Kolhe S (2023) A practical approach towards utilisation of the net-shaped micro-structures developed in the lifting plate Hele-Shaw cell for micro-mixing. Int J Interact Des Manuf IJIDeM. https://doi.org/10.1007/s12008-022-01167-7
4. Kale BS, Bhole K (2019) Parametric analysis for forming meso fractals from nanoparticle seeded resin in Hele Shaw cell. IOP Conf Ser Mater Sci Eng 577(1):012154. https://doi.org/10.1088/1757-899X/577/1/012154
5. Gandhi P, Bhole K (2013) Characterization of 'bulk lithography' process for fabrication of three-dimensional microstructures. J Micro Nano-Manuf 1(4). https://doi.org/10.1115/1.4025461
6. Bhole KS, Kale B (2022) Techniques to minimise stair-stepping effect in micro-stereolithography process: a review. Adv Mater Process Technol 8(4):3615–3634. https://doi.org/10.1080/2374068X.2021.1970997
7. Singare AA, Kale BS, Bhole KS (2018) Experimental characterization of meso-micro fractals from nanoparticle seeded resin in lifting plate hele-shaw cell. Mater. Today Proc 5(11), Part 3:24213–24220. https://doi.org/10.1016/j.matpr.2018.10.216
8. Kale BS, Bhole KS, Sharma C (2022) Effect of anisotropies in formation of viscous fingering in lifting plate Hele-Shaw cell. Adv Mater Process Technol 8(4):3780–3793. https://doi.org/10.1080/2374068X.2021.2013679
9. T. ul Islam and P. S. Gandhi, "Spontaneous Fabrication of Three-Dimensional Multiscale Fractal Structures Using Hele-Shaw Cell," *J. Manuf. Sci. Eng.*, vol. 139, no. 3, Oct. 2016, doi: https://doi.org/10.1115/1.4034624.
10. Sinha S, Dutta T, Tarafdar S (2008) Adhesion and fingering in the lifting Hele-Shaw cell: Role of the substrate. Eur Phys J E 25(3):267–275. https://doi.org/10.1140/epje/i2007-10289-9
11. Zare Vamerzani B, Zadehkabir A, Saffari H, Hosseinalipoor SM, Mazinani P, Honari P (2021) Experimental analysis of fluid displacement and viscous fingering instability in fractured porous medium: effect of injection rate. J Braz Soc Mech Sci Eng 43(2):66. https://doi.org/10.1007/s40430-020-02790-9
12. ul Islam T, Gandhi PS (2018) Controlling interfacial flow instability via micro engineered surfaces towards multiscale channel fabrication. In :ASME 2018 16th international conference on nanochannels, microchannels, and minichannels. https://doi.org/10.1115/ICNMM2018-7668
13. Kasambe PV, Bhole KS, Bage AA, Raykar NR, Bhoir DV (2022) Analytical modeling and numerical investigation of a variable width piezoresistive multilayer polymer micro-cantilever

air flow sensor. Adv. Mater. Process. Technol. 8(4):4365–4383. https://doi.org/10.1080/2374068X.2022.2076974
14. Saffman PG (1986) Viscous fingering in Hele-Shaw cells. J Fluid Mech 173:73–94. https://doi.org/10.1017/S0022112086001088
15. Kessler DA, Levine H (1986) Coalescence of Saffman-Taylor fingers: a new global instability. Phys Rev A 33(5):3625–3627. https://doi.org/10.1103/PhysRevA.33.3625
16. ul Islam T, Gandhi PS (2017) Viscous fingering in multiport hele shaw cell for controlled shaping of fluids. Sci Rep 7(1):16602. https://doi.org/10.1038/s41598-017-16830-3
17. Kale BS, Bhole K, Khond P (2019) Experimental modeling of meso fractals generated from non newtonian fluid from lifting plate Hele-Shaw Cell 9(2):29–35 (2019)

A Comprehensive Overview on Additive Manufacturing Processes: Materials, Applications, and Challenges

Nikhil Bharat, Rajat Jain, and P. S. C. Bose

1 Introduction

The term "additive manufacturing" refers to the process of using a digital 3D model as a starting point for the creation of a physical object. The process requires layering elements in successive prints. First came stereolithography (SLA), then came advances like contour crafting, inkjet printing, fused deposition modeling, and powder bed fusion. Three-dimensional printing (3D printing), which makes use of a number of distinct processes, materials, and tools, has developed through time and has the potential to revolutionize manufacturing and supply chain management. Industries such as prototyping, construction, and biomechanical engineering all benefit from additive manufacturing. Slow and restricted progress toward automation, less waste, and other benefits of 3D printing in the construction sector [1–5].

The development of new additive manufacturing technologies and materials always leads to the discovery of new applications. The expiry of past copyrights is one of the primary reasons why this ability is becoming more accessible. This has enabled fabricators to choose to invest in novel 3D printing equipment, which is one of the primary reasons why this skill is becoming more accessible. As a consequence of recent technological advancements, the cost of 3D printers has decreased, making it possible for these machines to be used in a wider variety of settings, including labs, libraries, private homes, and even educational institutions. The ability of 3D printing to create prototypes in a short amount of time and at a low cost first attracted a lot of attention from designers and engineers, who used it to make prototypes that were both aesthetically pleasing and practically useful [6, 7].

Printing in three dimensions presents a number of difficulties and constraints, including the lack of stiffness and strength in the printed construction materials, as well as the dimensions of the printed objects themselves. The fact that the suggested

N. Bharat (✉) · R. Jain · P. S. C. Bose
Department of Mechanical Engineering, National Institute Technology, Warangal 506004, India
e-mail: nikhilbharat@student.nitw.ac.in

V. S. Sharma et al. (eds.), *Machining and Additive Manufacturing*, Lecture Notes in Mechanical Engineering, https://doi.org/10.1007/978-981-99-6094-1_10

printed magnitude does not fit the criteria of fully 3D printers is an additional issue. Each individual printer has its own set of structural element production parameters, thus the proposed printed magnitude does not match these standards. Issues have also been expressed about the technology used for 3D printing, such as the demand for support to sustain the load of the object that is being made until the substance becomes sufficiently strong. Mix design and the use of coarse particles provide another challenge for the 3D printing of concrete. Inside the confines of their 3D printing software [8, 9].

The development of 3D printing technology may be traced back to the production of 3D assemblies one layer at a time, layer by layer, directly from computer-aided design (CAD) models. Printing in three dimensions (3D printing) is a cutting-edge technique that has shown to be incredibly versatile. It enables a variety of options and expands the scope of activities that are aimed at boosting industrial productivity. Nowadays, 3D printing technology can be used to create objects out of a wide variety of materials, including the more common thermoplastics, metals, graphene-based materials, and ceramics [10–13].

Being a relatively new technology, 3D printing has the potential to significantly alter manufacturing and assembly lines. Thanks to 3D printing, production costs will be minimal and quantity requirements will be easily met. More weight will be given to customer input throughout manufacturing. Customers may specify their own requirements for the final product and have it made accordingly. Concurrently, 3D printing facilities will be located closer to clients, allowing for a more adaptable and accessible industrial process and better quality control. In addition, the equipment needed for 3D printing reduces the requirement for transport across borders. This is because it is more efficient to use fleet monitoring technology to complete all supplies when production facilities are situated closer to the final destination. Furthermore, the company's logistics may be affected by the incorporation of 3D printing equipment. Companies' internal logistics divisions may oversee the whole journey and provide comprehensive, end-to-end services [14].

2 Additive Manufacturing Processes

2.1 Extrusion-Based Process

Extrusion-based additive manufacturing systems use a nozzle or extruder to deposit molten material layer by layer to create a 3D object. These systems typically use thermoplastic materials such as ABS or PLA, which are fed through a heated nozzle and deposited onto a build platform in the desired pattern. The process begins with a 3D model, which is sliced into thin layers using specialized software. The extruder moves along the X and Y axes, depositing the molten material in the desired pattern. As the material cools, it solidifies and fuses with the previous layer to create a strong and durable final part. Extrusion-based systems are widely used in additive manufacturing due to their versatility and relatively low cost. They are capable of producing a wide range of objects, from simple geometric shapes to highly complex designs. Additionally, these systems can use a variety of materials, allowing for a range of functional and aesthetic properties in the final part. However, extrusion-based systems can have some limitations in terms of surface finish, dimensional accuracy, and strength, which can be influenced by factors such as the type of material used, the layer height, and the printing speed. As a result, careful design and processing considerations are required to produce high-quality parts using extrusion-based systems [15]. Figure 1 shows the extrusion-based additive manufacturing process.

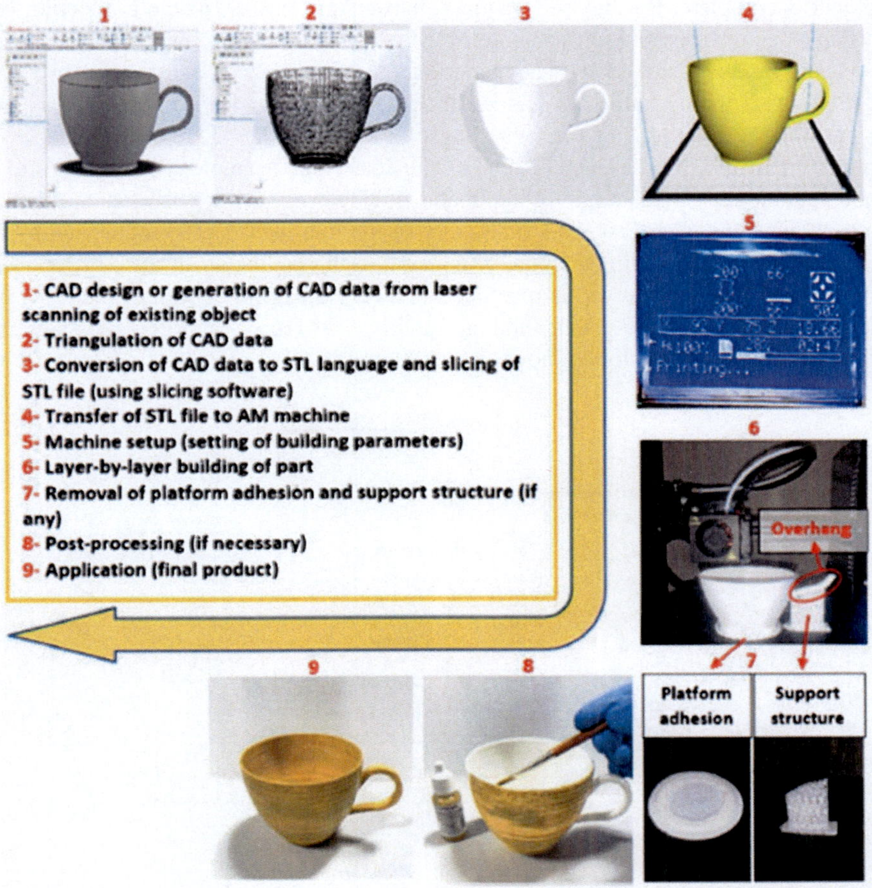

Fig. 1 Standardized additive manufacturing method based on extrusion, starting with CAD renderings [15]

2.2 Powder Bed Fusion Process

Powder bed fusion is an additive manufacturing process that creates 3D objects layer-by-layer by selectively fusing together particles of a powdered material using a heat source, such as a laser or electron beam. There are several types of powder bed fusion processes, including selective laser sintering (SLS), selective laser melting (SLM), and electron beam melting (EBM). In all powder bed fusion processes, a thin layer of powder is spread over a build platform and a heat source is used to selectively fuse together the particles in the desired pattern. Once the layer is complete, another layer of powder is spread over the top and the process is repeated until the object is complete. Powder bed fusion is particularly useful for the production of complex geometries and intricate internal structures, as it can create parts with high levels of accuracy and detail. The technology is used in a wide range of industries, including aerospace, automotive, medical, and consumer products. One of the advantages of powder bed fusion is that it allows for the use of a wide range of materials, including metals, plastics, and ceramics. This makes the technology highly versatile and suitable for a wide range of applications. However, powder bed fusion can be a time-consuming process, and the quality of the final part can be influenced by a range of factors, including the type of material used, the quality of the powder, and the operating conditions of the machine. As a result, careful control of the process is required to ensure consistent and high-quality parts [16]. The systematic diagram of laser powder bed fusion is shown in Fig. 2.

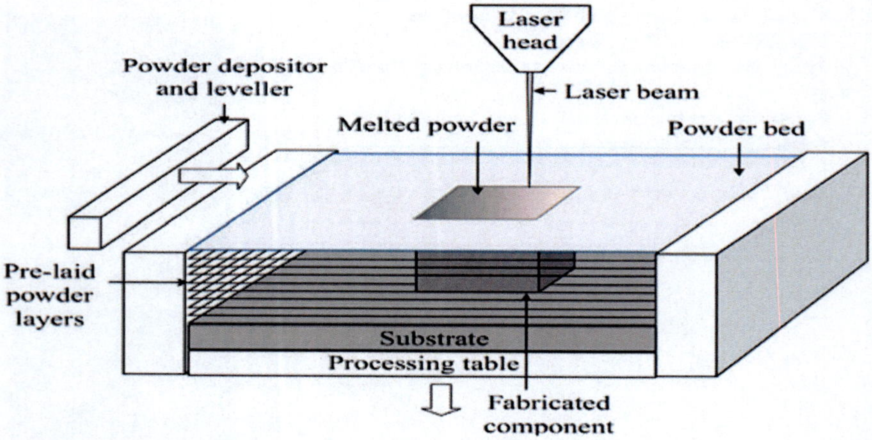

Fig. 2 Laser powder bed fusion (LPBF) technique [17]

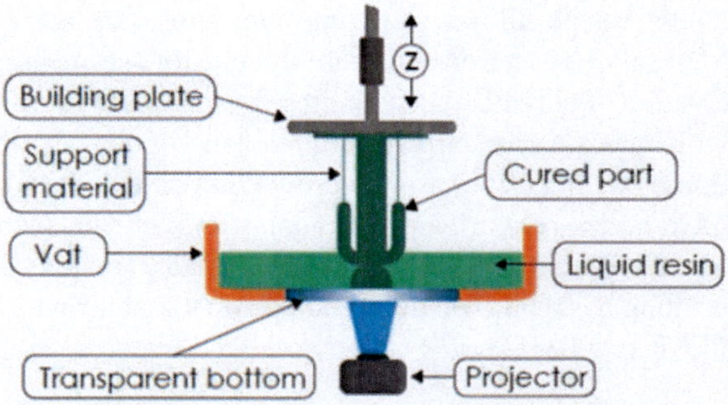

Fig. 3 Photopolymerization process [19]

2.3 Photopolymerization Process

Photopolymerization is an additive manufacturing process that uses light to selectively cure and solidify a liquid resin into a 3D object. In this process, a liquid photopolymer resin is selectively exposed to a light source, typically an ultraviolet (UV) laser or projector, in the desired pattern to cure the material and create the object layer-by-layer. In photopolymerization, the liquid resin is typically contained in a vat, and the build platform is lowered into the resin. The light source then cures the material in the desired pattern, solidifying the resin into a layer of the object. The build platform is then lifted out of the resin and a new layer of liquid resin is spread over the top, and the process is repeated until the object is complete. One of the advantages of photopolymerization is that it can create highly detailed and accurate parts with smooth surface finishes. The process is particularly useful for creating objects with intricate internal structures or fine details, such as dental models, jewelry, and small mechanical components. However, photopolymerization can have some limitations, such as the need for support structures to prevent the object from collapsing during printing, which can be time-consuming and require post-processing to remove. Additionally, the materials used in photopolymerization can be more expensive than those used in other types of additive manufacturing, making them less suitable for large-scale production [18]. The schematic diagram of photopolymerization process is shown in Fig. 3.

3 AM Materials for 3D Printing

3.1 Metals and Alloys

The benefits of metal 3D printing have aroused the interest of the automotive, aerospace, industrial, and medical sectors. Metals are highly desirable due to their adaptability and versatility; they can be used for everything from 3D printing to making airplane components to producing human organs. Titanium alloys, nickel alloys, cobalt alloys, and aluminum alloys are a few examples of such materials. Metal alloys based on cobalt are suitable for use in 3D printed dental applications. The creation of polymer parts, ranging from simple patterns to fully functional assemblies with intricate geometry, is a common use of the 3D printing technique. Fused deposition modeling allows for the production of 3D printed objects by depositing consecutive layers of extruded thermoplastic filaments such as polyethylene, polypropylene, acrylonitrile butadiene styrene (ABS), and polylactic acid (PLA). Recently, 3D printing has made use of thermoplastic filaments with higher melting points, such as PEEK and PMMA [20–24].

3.2 *Composite Materials*

Composites combine materials with distinct physical and chemical characteristics. Composites offer superior mechanical, physical, and chemical qualities. 3D printers can print composites, which are used in additive manufacturing owing to their cheap cost, availability, improved characteristics, and cost-effective product customization with great precision compared to conventional techniques. Composites are used in manufacturing, electronics, aircraft, biomedical, structural, sports, wearable, and cars. Recent research has succeeded in fabricating a wide range of composites, including particle-reinforced composites, natural and synthetic fiber-reinforced composites, nanomaterial-reinforced composites, ionic polymer metal composites, graphene-based composites, and lightweight cellular composites. These categories include several composite materials. Fused deposition modeling (FDM), and selective laser sintering (SLS) may be used to print composite materials [25–28].

3.3 *Biomaterials*

Healthcare items are made using biomaterials. As 3D printing becomes cheaper, this industry relies on additive manufacturing. Recent research suggests that 3D printing advances like enhanced accuracy, speed, capacity to print complicated structures, decreased cost and material waste, and more promote the usage of bio-materials in additive manufacturing. Additive manufacturing technology enables medical equipment, cell printing, and implant materials to progress greatly. 3D printing can directly build defective sites from patient CT and MRI data. First, capture the region of interest from medical imaging. Then, create a 3D geometry from the dataset unique to the area of interest. After that, transform the 3D object into a file suited for printing. Finally, pick a 3D printer and material. 3D printing technology is employed in tissue engineering, organ model printing, customized scaffold printing, and dentistry. Additive manufacturing may employ many biomaterials. Biological materials such as living cells, extracellular matrix (ECM), and tissue-specific biomaterials are used in additive manufacturing for the fabrication of tissue constructs for regenerative medicine applications. These materials are typically used in bioprinting, which is a type of additive manufacturing that involves the printing of living cells and biomaterials to create functional tissue constructs [29–33].

4 Applications and Challenges in Additive Manufacturing

4.1 Applications of Additive Manufacturing

Aerospace: The aerospace industry is one of the primary users of additive manufacturing. Additive manufacturing is used to manufacture components with complex geometries that are difficult to produce using traditional methods. The parts produced using additive manufacturing are lighter and have improved mechanical properties, which leads to more fuel-efficient and high-performance aircraft.

Biomedical: Additive manufacturing has revolutionized the field of biomedical engineering. The technology is used to produce custom implants, prosthetics, and other medical devices. The technology has the potential to transform healthcare, enabling faster, cheaper, and more personalized medical treatments.

Automotive: The automotive industry is another major user of additive manufacturing. The technology is used to produce lightweight components with improved strength and durability. Additive manufacturing is also used to produce complex geometries, such as engine parts, that are difficult or impossible to produce using traditional manufacturing methods.

Architecture: Additive manufacturing is used in architecture to produce complex models and prototypes of buildings. The technology enables architects to create intricate designs that would be difficult to produce using traditional methods.

Consumer Goods: Additive manufacturing is also used in the consumer goods industry to produce customized products, such as jewelry, toys, and household items. The technology enables manufacturers to produce products with a high level of customization, which leads to improved customer satisfaction.

4.2 Challenges in Additive Manufacturing

Material Selection: The range of materials that can be used in additive manufacturing is limited compared to traditional manufacturing processes. Therefore, it can be challenging to find the right material for a specific application.

Part Quality: The quality of the parts produced using additive manufacturing is highly dependent on the printing parameters used, such as layer thickness, print speed, and temperature. Therefore, optimizing the printing parameters is essential to produce parts with consistent quality.

Cost: Additive manufacturing can be more expensive than traditional manufacturing methods, particularly when producing large quantities of parts. However, the cost of additive manufacturing has been decreasing in recent years, and the technology has become more affordable.

Post-processing: Additive manufacturing often requires post-processing, such as sanding, polishing, or painting, to achieve the desired surface finish. Post-processing can add time and cost to the manufacturing process.

Intellectual Property: Additive manufacturing has raised concerns about intellectual property. The technology enables the production of parts without the need for traditional manufacturing tooling, which makes it difficult to protect intellectual property rights.

5 Conclusion

This review covers industrial 3D printing applications. 3D printing is entering industrial engineering, and it helps people, companies, and governments. So, better 3D printing tactics need more data. Businesses and governments may enhance their 3D printing setup with more proof. So, this article describes several 3D printing methods, manufacturing components, and applications. Further investigation will examine the various 3D printing machines and their resources. Manufacturing advances need innovative research in materials, processes, and product design. Product complexity requires new and imaginative manufacturing methods. Additive manufacturing is a contemporary industrial trend because of its benefits and challenges. Researchers analyzed and reviewed. This article presents the results of an in-depth investigation on AM that was carried out. The importance of component alignment, construction time estimate, and budget calculation has been explored. The most important part of this study is identifying AM approach issues. AM difficulties include part size constraints, anisotropic mechanical characteristics, high costs, low accuracy, mass manufacturing, wrapping, layer misalignment, and material limitations.

6 Future Scope

Due to its widespread use in aerospace, automotive, medical, and consumer goods sectors, additive manufacturing (AM), commonly known as 3D printing, is a fast-developing technology. Despite substantial advances in AM technology, some hurdles remain to fully fulfil its promise.

The creation of new materials is one topic that might be investigated. Advanced polymers, composites, and metals are just some of the novel materials that have attracted a lot of attention for application in additive manufacturing. Possible directions for further study include expanding our understanding of the capabilities of novel materials and designing materials that are more amenable to AM processes.

Researchers may, for instance, create materials with enhanced attributes like strength, durability, and heat resistance, or with specialized features like conductivity or biocompatibility. Research into better methods of process control and quality assurance is also vital. In order to guarantee that the components made using AM methods are up to par, it will be necessary to refine process control and quality assurance methods. This is especially significant in sectors where high-quality components are crucial, such as the aerospace and medical fields. Better process control may also allow for more dependable printing at higher speeds, saving time and money in production. The use of AM in mass manufacturing is another promising area of research. While AM methods work best for low-volume manufacturing, there is much enthusiasm for scaling up to factory-scale production. The potential for AM to be employed in the industrial manufacturing of large quantities of components may motivate further study. Researchers may, for instance, create faster and more efficient printing methods to facilitate mass manufacturing. The ability to print using more than one material is also quite intriguing. As several materials may be used in the printing process, more complicated components with enhanced functionality may be manufactured. The ability to print multi-material components with fine-grained control over material location and composition may be the focus of future studies. Last but not least, sustainability is a focus of much study in AM. Although, while AM is generally regarded as a more environmentally friendly manufacturing technology than more conventional approaches, it may always do better. Improved recyclability, less waste, and lower energy use are all potential areas of investigation for the near future. New materials derived from sustainable sources might be developed, or methods of recycling or reusing AM components could be investigated.

References

1. Ru C, Luo J, Xie S, Sun Y (2014) A review of non-contact micro- and nano-printing technologies. J Micromech Microeng 24:53001. https://doi.org/10.1088/0960-1317/24/5/053001
2. Bhushan B, Caspers M (2017) An overview of additive manufacturing (3D printing) for microfabrication. Microsyst Technol 23:1117–1124. https://doi.org/10.1007/s00542-017-3342-8
3. Jadhav A, Jadhav VS (2022) A review on 3D printing: an additive manufacturing technology. Mater Today Proc 62:2094–2099. https://doi.org/10.1016/j.matpr.2022.02.558
4. Mahamood RM, Jen T-C, Akinlabi SA, Hassan S, Akinlabi ET (2023) Chapter 1—Introduction to additive manufacturing technologies. In: Kumar A, Mittal RK, Haleem ABT-A in AM (eds) Additive manufacturing materials and technologies. Elsevier, pp 3–13
5. Zhang X, Liou F (2021) Chapter 1—Introduction to additive manufacturing. In: Pou J, Riveiro A, Davim JPBT-AM (eds) Handbooks in advanced manufacturing. Elsevier, pp 1–31
6. Chohan JS, Singh R, Boparai KS, Penna R, Fraternali F (2017) Dimensional accuracy analysis of coupled fused deposition modeling and vapour smoothing operations for biomedical applications. Compos Part B Eng 117:138–149. https://doi.org/10.1016/j.compositesb.2017.02.045
7. Parandoush P, Lin D (2017) A review on additive manufacturing of polymer-fiber composites. Compos Struct 182:36–53. https://doi.org/10.1016/j.compstruct.2017.08.088

8. Travitzky N, Bonet A, Dermeik B, Fey T, Filbert-Demut I, Schlier L, Schlordt T, Greil P (2014) Additive manufacturing of ceramic-based materials. Adv Eng Mater 16:729–754. https://doi.org/10.1002/adem.201400097
9. Dou R, Wang T, Guo Y, Derby B (2011) Ink-jet printing of zirconia: Coffee staining and line stability. J Am Ceram Soc 94:3787–3792. https://doi.org/10.1111/j.1551-2916.2011.04697.x
10. Khoshnevis B (2004) Automated construction by contour crafting—related robotics and information technologies. Autom Constr 13:5–19. https://doi.org/10.1016/j.autcon.2003.08.012
11. Singh R, Singh N, Amendola A, Fraternali F (2017) On the wear properties of $Nylon_6$-SiC-Al_2O_3 based fused deposition modelling feed stock filament. Compos Part B Eng 119:125–131. https://doi.org/10.1016/j.compositesb.2017.03.042
12. Regassa Hunde B, Debebe Woldeyohannes A (2022) Future prospects of computer-aided design (CAD)—a review from the perspective of artificial intelligence (AI), extended reality, and 3D printing. Results Eng 14:100478. https://doi.org/10.1016/j.rineng.2022.100478
13. Scerrati A, Trovalusci F, Albanese A, Ponticelli GS, Tagliaferri V, Sturiale CL, Cavallo MA, Marchese E (2019) A workflow to generate physical 3D models of cerebral aneurysms applying open source freeware for CAD modeling and 3D printing. Interdiscip. Neurosurg. 17:1–6. https://doi.org/10.1016/j.inat.2019.02.009
14. Han Q, Geng Y, Setchi R, Lacan F, Gu D, Evans SL (2017) Macro and nanoscale wear behaviour of Al-Al_2O_3 nanocomposites fabricated by selective laser melting. Compos Part B Eng 127:26–35. https://doi.org/10.1016/j.compositesb.2017.06.026
15. Altıparmak SC, Yardley VA, Shi Z, Lin J (2022) Extrusion-based additive manufacturing technologies: state of the art and future perspectives. J Manuf Process 83:607–636. https://doi.org/10.1016/j.jmapro.2022.09.032
16. Abd-Elaziem W, Elkatatny S, Abd-Elaziem A-E, Khedr M, Abd El-baky MA, Hassan MA, Abu-Okail M, Mohammed M, Järvenpää A, Allam T, Hamada A (2022) On the current research progress of metallic materials fabricated by laser powder bed fusion process: a review. J Mater Res Technol 20:681–707. https://doi.org/10.1016/j.jmrt.2022.07.085
17. Tian Z, Zhang C, Wang D, Liu W, Fang X, Wellmann D, Zhao Y, Tian Y (2020) A review on laser powder bed fusion of inconel 625 nickel-based alloy
18. Andjela L, Abdurahmanovich VM, Vladimirovna SN, Mikhailovna GI, Yurievich DD, Alekseevna MY (2022) A review on Vat Photopolymerization 3D-printing processes for dental application. Dent Mater 38:e284–e296. https://doi.org/10.1016/j.dental.2022.09.005
19. Davoudinejad A, Diaz Perez LC, Quagliotti D, Pedersen DB, Albajez García JA, Yagüe-Fabra JA, Tosello G (2018) Geometric and feature size design effect on vat photopolymerization micro additively manufactured surface features. In: Proc. - 2018 ASPE euspen Summer Top. Meet. Adv. Precis. Addit. Manuf. pp 55–59
20. Duda T, Raghavan LV (2016) 3D Metal Printing Technology. IFAC-PapersOnLine. 49:103–110. https://doi.org/10.1016/j.ifacol.2016.11.111
21. Lee T, Jeong W, Chung S, So KP, Ryu HJ (2023) Novel solid-state metal powder surface modification process for additive manufacturing of metal matrix composites and alloys. Appl Surf Sci 615:156364. https://doi.org/10.1016/j.apsusc.2023.156364
22. Wei F, Cheng B, Chew LT, Lee JJ, Cheong KH, Wu J, Zhu Q, Tan CC (2022) Grain distribution characteristics and effect of diverse size distribution on the Hall-Petch relationship for additively manufactured metal alloys. J Mater Res Technol 20:4130–4136. https://doi.org/10.1016/j.jmrt.2022.09.006
23. Alkhatib SE, Sercombe TB (2022) High strain-rate response of additively manufactured light metal alloys. Mater Des 217:110664. https://doi.org/10.1016/j.matdes.2022.110664
24. Moghimian P, Poirié T, Habibnejad-Korayem M, Zavala JA, Kroeger J, Marion F, Larouche F (2021) Metal powders in additive manufacturing: a review on reusability and recyclability of common titanium, nickel and aluminum alloys. Addit Manuf 43:102017. https://doi.org/10.1016/j.addma.2021.102017
25. Jiang L, Peng X, Walczyk D (2020) 3D printing of biofiber-reinforced composites and their mechanical properties: a review. Rapid Prototyp J 26:1113–1129. https://doi.org/10.1108/RPJ-08-2019-0214

26. Adeniran O, Cong W, Aremu A (2022) Material design factors in the additive manufacturing of carbon fiber reinforced plastic composites: a state-of-the-art review. Adv Ind Manuf Eng 5:100100. https://doi.org/10.1016/j.aime.2022.100100
27. Yao D, Wang J, Cai Y, Zhao T, An X, Zhang H, Fu H, Yang X, Zou Q, Wang L (2022) Composition regulation of composite materials in laser powder bed fusion additive manufacturing. Powder Technol 408:117795. https://doi.org/10.1016/j.powtec.2022.117795
28. McCarthy E, Brabazon D (2021) Additive manufacturing for sustainability of composite materials production. Presented at the (2021)
29. Vyavahare S, Mahesh V, Mahesh V, Harursampath D (2023) Additively manufactured metabiomaterials: a state-of-the-art review. Compos Struct 305:116491. https://doi.org/10.1016/j.compstruct.2022.116491
30. Singh D, Kumar R, Ramniwas S (2023) Chapter 6—from drosophila material to functional structures: biomimetic through additive manufacturing technology. In: Singh S, Prakash C, Ramakrishna SBT-IP, M in AM (eds) Woodhead publishing reviews: mechanical engineering series. Woodhead Publishing, pp 129–151
31. Davoodi E, Montazerian H, Mirhakimi AS, Zhianmanesh M, Ibhadode O, Shahabad SI, Esmaeilizadeh R, Sarikhani E, Toorandaz S, Sarabi SA, Nasiri R, Zhu Y, Kadkhodapour J, Li B, Khademhosseini A, Toyserkani E (2022) Additively manufactured metallic biomaterials. Bioact Mater 15:214–249. https://doi.org/10.1016/j.bioactmat.2021.12.027
32. Pesode P, Barve S (2022) Additive manufacturing of metallic biomaterials and its biocompatibility. Mater Today Proc. https://doi.org/10.1016/j.matpr.2022.11.248
33. Simorgh S, Alasvand N, Khodadadi M, Ghobadi F, Malekzadeh Kebria M, Brouki Milan P, Kargozar S, Baino F, Mobasheri A, Mozafari M (2022) Additive manufacturing of bioactive glass biomaterials. Methods 208:75–91. https://doi.org/10.1016/j.ymeth.2022.10.010

An Overview of the Untapped Potential of Soft Robotic Arms with Integration of Machining Tools

Shaarif Hamdule and Shashikant S. Goilkar

1 Introduction

Soft robotics is a branch of robotics that utilizes actuators made of flexible materials instead of rigid links, with or without minimal number of rigid parts for mounting purposes. These actuators are inspired by nature, movements of an elephant's trunk [1], locomotion of earthworms [2], jellyfish, and rays [3] are a few examples of biomimetics in this field. Actuation uses pneumatical control, electrical activation, on-board combustion of fuel [4], electrothermal actuation for shape-memory alloys, optical actuation [5], and biohybrid actuation [6]. Soft robots find their applications in various fields. In industries, soft grippers are used to handle fragile objects with minor uncertainties in shapes and sizes. In the medical field, they are used as wearables [7], to augment the motion capabilities of a healthy person on those suffering from muscle weakness due to physical or neurological disorders, and as prosthetics [8] as a cheap and lightweight solution. Other uses of soft robotic arms (SRA) include deep sea exploration [9] of delicate organisms, nuclear decontamination [10], and much more. Several other applications such as planetary exploration [11–13], and surgical uses [14] are actively studied.

1.1 Soft Actuators and Sensors

In rigid robots, actuators, often paired with encoders, at joints, position the links in desirable orientation. In soft robotics, the actuation process is integrated into and throughout the manipulator. SRA employing bellows and Honeycomb Pneumatic Networks (HPN) [15] configure themselves due to differences in the inflation state

S. Hamdule (✉) · S. S. Goilkar
Finolex Academy of Management and Technology, Ratnagiri, MH 415639, India
e-mail: m.shaarif@gmail.com

V. S. Sharma et al. (eds.), *Machining and Additive Manufacturing*, Lecture Notes in Mechanical Engineering, https://doi.org/10.1007/978-981-99-6094-1_11

of one unit of manipulator with respect to the inflation state of other units. That is, inflated bellows or HPN tends to take a convex profile over the uninflated units. McKibben muscles and other fiber-reinforced actuators [16] extend, contract, bend, or twist depending on the configurations of fiber covering the inflatable bladder. As non-pneumatic actuators are not usually implemented in SRA, they are omitted in this paper.

To implement a closed-loop control system, the use of encoders is imperative. Similar to actuators, these encoders are made of flexible material too. Eutectic Ga-In sensor [17] uses liquid metals encased in silicones. When the setup is stretched, dimensional changes of channels containing the liquid alloy affect its impedance, which can be calibrated to the measure of change in its length. Other such sensors, that rely on impedance are the smart braid sensor [18] and the silicone-textile composite capacitive strain sensor [19] with the former relying on change in inductance and resistance and the later, relying mostly on change in capacitance.

1.2 Benefits and Constrains of Soft Robots

Designing a conventional robot requires taking spatial constraints into account such as the surrounding machinery and processes, as they define the dimensional characteristics. To carry out any operation, the automaton changes the position and orientation of its links between a number of pre-set states. Any obstructions amidst the paths of transition may hinder the operation. In the case of SRA, links cannot only attain various lengths but also curve along these lengths, bending across the obstruction. Hence, soft robots are said to be highly compliant to their surroundings and can be used in unspecified environments. This merit is not only limited to arms but also grippers, as they easily conform to objects that are fragile or have minor uncertainties in shapes and sizes. Soft robots also provide a safer working environment, as they are mostly made of flexible materials with minimal amounts of rigid components. This makes soft robots ideal in places with significant human–machine interaction such as personal care for elderly [20], or in medical field for endoscopy [21].

Soft robots come with their own set of complications. Actuators are highly influenced by the effect of gravity. Unlike rigid links, soft actuators may undergo sagging along the length and the desired position of the end effector may not be attained. Factors such as extension, contraction, bending, and torsion of links come into play. Actuators are also application-specific and, therefore, a general control strategy cannot provide solution in all scenarios, making a design of control strategies [22], a major problem. However, with the recent advancements in machine learning and modern computational power, the solution gets more viable by the day.

2 Present Scenario of Soft Robotic Arms

To facilitate the ease of operation in unspecified environments, SRA can be used to provide better compliance to the surroundings by bending around obstructions amidst the path to the final orientation. SRA resembles the movements of that of an elephant's trunk. A rigid alternative that mimics such motion characteristics would be a hyper-redundant manipulator, having multitudes of joints and therefore, a myriad of actuators and encoders.

Bionic Handling Assistant [23] developed by Festo in cooperation with Fraunhofer IPA comprises three modules for spatial movements, as well as a hand axis and a gripper. Each basic module comprises of three circularly arranged actuators with a loop-like design, tapering at an angle of 3°. Compressed air is used to inflate actuators and orient the arm in the desired position, while, resetting is done because of the loop-like structure that acts as a spring when the air is discharged. Bowden cable potentiometers are used to measure the extension and control the arm's spatial movement. The hand axis consists of three actuators, that are arranged around a ball joint. The gripper uses adaptive fingers to conform to the object. In total, the arm provides 11 degrees of freedom and extends a total of 40 cm in length. The entirety of the arm was constructed using selective laser sintering, hence giving high flexibility to the arm with lightweight design.

The OctArm [24] is another pneumatically actuated SRA that makes use of McKibben actuators, constructed by covering latex tubing with double helical weave, and plastic mesh sheaths. These extensors provide a large strength-to-weight ratio. OctArm V is tested to successfully grasp objects of different shapes and sizes, even maintaining the stability of grasp under dynamic disturbances. Using 8.27 bar of pressure, this model is said to have a vertical load-carrying capacity of 890 N and a transverse load-carrying capacity of 250 N at its proximal section.

The Tentacle Manipulator developed by Kinetic Sciences Inc. (KSI) is driven by a hybrid system of pneumatic bellows and electric motors. In addition to extending and contracting, it can bend independently in two or more regions along its length. Contraction and bending are achieved with the use of tendons threading cable guides, providing a total of 6 degrees of freedom. Kinetic Sciences Inc. has sold a Tentacle to Battelle Pacific Northwest Labs at the Department of Energy, Hanford site in Washington State for application in clearing nuclear decontamination of a hot cell by vacuuming radioactive detritus off the floor.

All the aforementioned designs made use of pneumatic actuators in conjunction with tendon cables. Use of tendons can be explained with the demerits of using a hyper-redundant robot. Although, hyper-redundant system provides a better solution in case of rigid robotics when used in unspecified environments, it is more practical to move the actuators and encoders off the body, otherwise, each actuator will have to carry the load of all the proceeding actuators, encoders, links, and other auxiliaries. Moving all such driving and sensing parts to the base of the arm and transmitting motion and torque by tendons [25] can overcome the limitations.

HPN manipulator mentioned earlier does not make use of tendons. It is constructed by integrating pneumatic networks into compressed honeycomb structures, made up of elastomers. These hexagonal frames have airbags within them that inflate during actuation. Two rows of HPN units, stacked one above the other, can be inflated with the left row separately from the right, and top rows separately from the bottom. Because of the structure's resilience, the non-inflated side gains the concave profile during bending. When both chambers in left and right or top and bottom are inflated together, mere lengthwise elongation takes place. The compressed honeycomb structure provides a high elongation rate and crush resistance.

3 The Untapped Potential of Soft Robotic Arms

All the earlier-mentioned uses of SRA were specific to only picking up, grasping, and placing objects, with KSI Tentacle Manipulator as an exception. The range of applications can widely be increased by focusing on the variety of tools that can be mounted at the end effector. For better integration in industrial automation, surgical purposes, exploratory bots, and several other fields, it is imperative for SRA to be able to perform operations such as drilling, grinding, etc. Thereby, implying a need for research for tools to be mounted at the end effector and a viable way to drive these tools such that the flexible setup can endure the undesired wobble caused by them. This study specifically concerns with drilling operations performed by a soft robotic arm.

The literature reviewed so far shows that with an exception of HPN manipulator, each design can be divided along their lengths into segments, with each segment consisting of circularly arranged pneumatic actuators. The relative pressure within each actuator of a segment decides its spatial configuration. And, the spatial configuration of each segment decides the orientation of the entire SRA, hence, also the position of the end effector. Therefore, a similar design can be used for a soft robotic drilling arm (see Fig. 1) with segments consisting of circularly arranged pneumatic actuators such as bellows and tendon cables to help in contraction and bending.

As the distal segment is only required to linearly elongate to thrust the drill bit inside the workpiece, it can be exempted from the use of tendon cables. SRA being relatively flexible may not necessarily endure the wobble caused by the drive. If the motor is mounted at the tip of the SRA, under dynamic load, the design becomes highly unstable. Hence, the motor needs to be moved off the body of the arm, to its base. Motion and torque required by the tools can be transmitted by a flexible shaft. However, the length of the arm is variable, but, the length of the flexible shaft is not. Providing appropriate slack within each segment to compensate for the variation in arm's length can be a viable solution. Flexible shaft can be placed at the core of the arm between the circularly arranged actuators. Pneumatic connections are explained in Fig. 2.

For further stabilization, laminar jamming structures [26] consisting of layers of sandpaper enclosed in flat silicone tubes in an arrangement as shown in Fig. 3 (left)

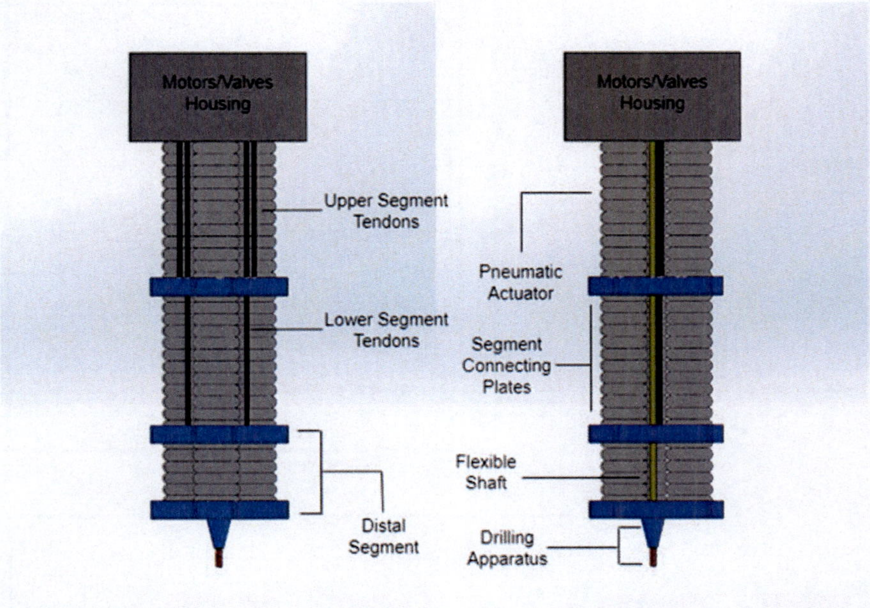

Fig. 1 Solid model of the proposed design

can be implemented in the form of a sleeve (solid model depicted in Fig. 3 (right)). The sleeve can jacket around all the modules except for the distal one. Once the SRA takes the final orientation, the sleeve can be subjected to vacuum, resulting in it getting stiffer, thereby, increasing the rigidity of the entire setup. During the drilling operation, the rigidity of the arm can be temporarily enhanced, and once the operation is completed, the setup can be reverted back to its compliant self.

The effectiveness of all the design decisions taken so far in order to stabilize the setup can be studied by measure of deviation from the required orientation at certain points of interest. A simple setup shown in Fig. 4 can help better study these deviations.

The coupling between adjacent segments can be mounted with sensors to study the stability provided by laminar jamming structures, similarly, displacement of end effector from its mean position can be studied for measuring the overall accuracy of the design.

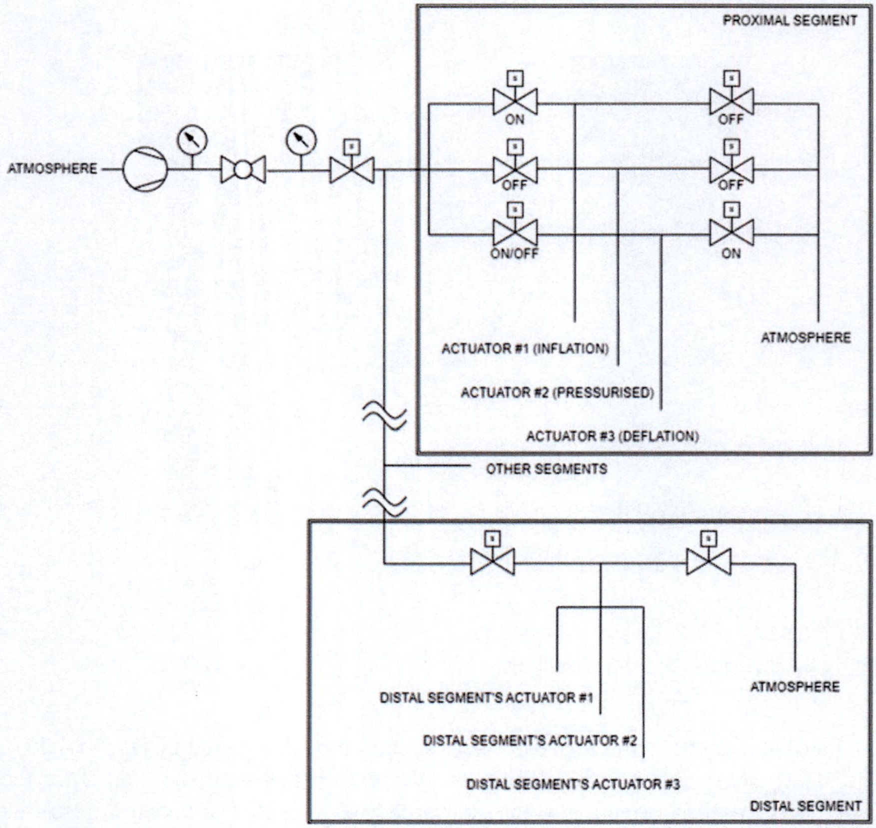

Fig. 2 Pneumatic circuit of the proposed design

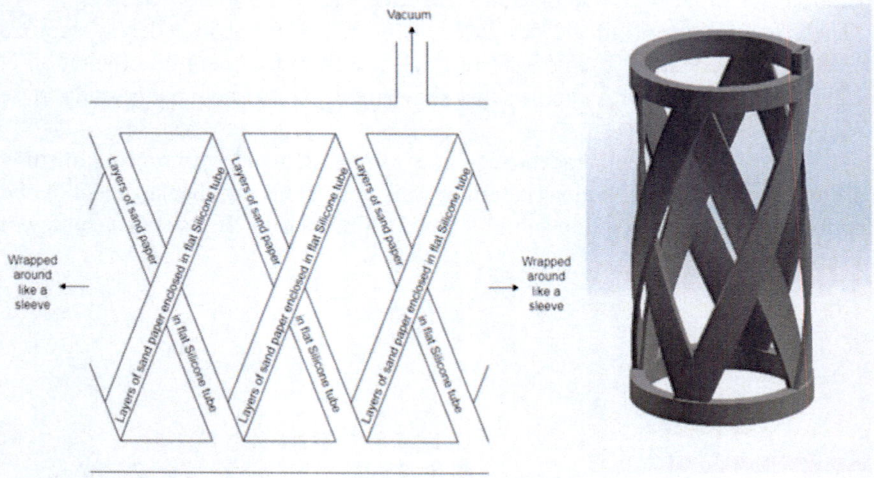

Fig. 3 Laminar jamming sleeve schematic (left) and solid model (right)

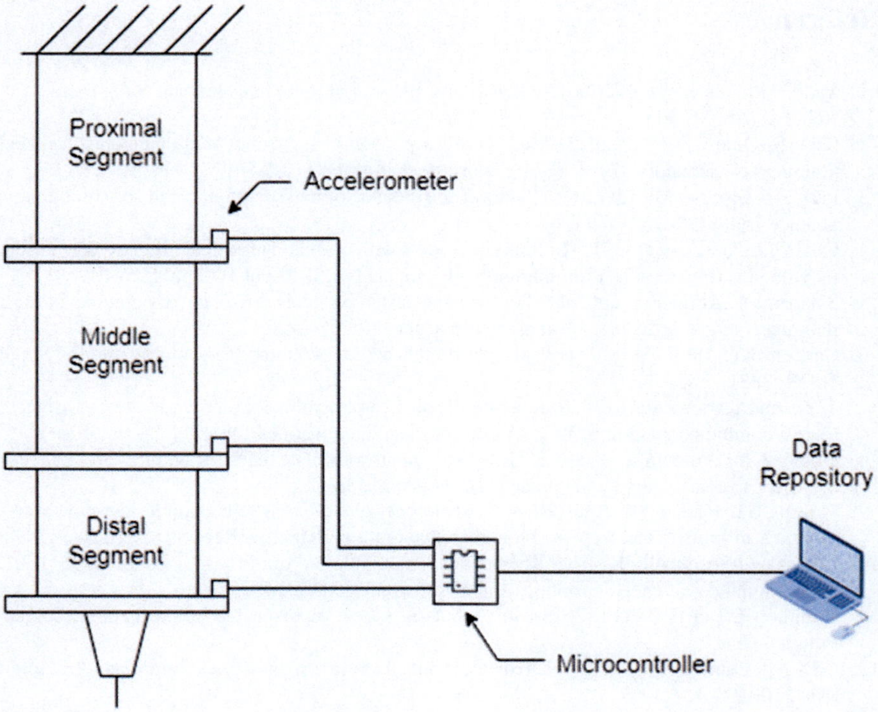

Fig. 4 Acquisition of data from soft robotic arm for testing

4 Conclusions

This study provides an overview on soft robotics arms, stating its present scenario as well as outlining the applications in research. Therefore, illuminating its untouched potential. Soft robotic arms are often limited to uses pertaining to grasping and relocation of objects. The range, however, can be further expanded by integrating tools at the end effector of the arm, thereby, making the arm capable of performing simple machining operations. The later part of this study states challenges in achieving these operations and proposes a design for a soft robotic drilling arm. Owing to the compliant nature to its environment and lightweight, such tool-integrated soft robotic arms can be used in a number of applications such as in the surgical field, for marine and space exploration missions, in industries dealing with products having complex frames with links often hindering the path of conventional robotic arms and many more.

References

1. Yang J, Pitarch EP et al (2006) Synthesis and analysis of a flexible elephant trunk robot. Adv Robot 20(6):631–59
2. Seok S, Onal CD et al (2013) Meshworm: a peristaltic soft robot with antagonistic nickel titanium coil actuators. IEEE/ASME Trans Mechatron 18(5):1485–97
3. Park S-J, Gazzola M et al (2016) Phototactic guidance of a tissue-engineered soft-robotic ray. Science 353(6295):158–162
4. Onal CD, Chen X et al (2017) Soft mobile robots with on-board chemical pressure generation. In: Robotics research-the 15th international symposium ISRR, vol 100. pp. 525–40
5. Satapathy P, Adiga R et al (2022) Porous nanocarbon particles drive large magnitude and fast photomechanical actuators. J Nanostructure Chem 12:235–248
6. Carlsen RW, Sitti M (2014) Bio-hybrid cell-based actuators for microsystems. Small 10(19):3831−3851
7. Thalman C, Artemiadis P (2020) A review of soft wearable robots that provide active assistance: trends, common actuation methods, fabrication, and applications. Wearable Technol 1:e3
8. Guoying G, Ningbin Z et al (2021) A soft neuroprosthetic hand providing simultaneous myoelectric control and tactile feedback. Nat Biomed Eng
9. Phillips BT, Becker KP et al (2018) A dexterous, glove-based teleoperable low-power soft robotic arm for delicate deep-sea biological exploration. Scientific Reports 8, Article 14779
10. Immega G, Antonelli K, The KSI tentacle manipulator. In: Proceedings of 1995 IEEE international conference on robotics and automation 3: 3149–54
11. Trimmer B, Lin H-T et al (2013) Soft robots in space: a perspective for soft robotics. Acta Futura 6:69–79
12. CSX Ng, Lum GZ (2022) Untethered soft robots for future planetary explorations? Adv Intell Syst 2100106
13. Methenitis G, Hennes D et al (2015) Novelty search for soft robotic space exploration. In: Proceedings of the 2015 annual conference on genetic and evolutionary computation (GECCO '15). pp 193–200
14. Runciman M, Darzi A et al (2019) Soft robotics in minimally invasive surgery. Soft Rob 6(4):423–443
15. Jiang H, Liu X et al (2016) Design and simulation analysis of a soft manipulator based on honeycomb pneumatic networks. In: 2016 IEEE international conference on robotics and biomimetics (ROBIO). pp 350–56
16. Daerden F, Lefeber D (2002) Pneumatic artificial muscles: actuators for robotics and automation. Eur J Mech Environ Eng 47(1):11–21
17. Ali MM, Narakathu BB et al (2016) Eutectic Ga-In liquid metal based flexible capacitive pressure sensor. In: 2016 IEEE SENSORS. pp 1–3
18. Felt W, Remy CD (2014) Smart braid: air muscles that measure force and displacement. In: 2014 IEEE/RSJ international conference on intelligent robots and systems. pp 2821–26
19. Atalay A, Sanchez V et al (2017) Batch fabrication of customizable silicone-textile composite capacitive strain sensors for human motion tracking. Adv Mater Technol 2(9):1700136
20. Manti M, Pratesi A et al (2016) Soft assistive robot for personal care of elderly people. In: 2016 6th IEEE international conference on biomedical robotics and biomechatronics (BioRob). pp 833–8
21. Gifari MW, Naghibi H et al (2019) A review on recent advances in soft surgical robots for endoscopic applications. Int J Med Robot Comput Assist Surg 15(5):e2010
22. George Thuruthel T, Ansari Y et al (2018) Control strategies for soft robotic manipulators: a survey. Soft Robot 5(2):149–63
23. Grzesiak A, Becker R et al (2011) The bionic handling assistant: a success story of additive manufacturing. Assem Autom 31(4):329–333
24. Grissom MD, Chitrakaran V et al (2006) Design and experimental testing of the OctArm soft robot manipulator. In: SPIE proceedings volume 6230, unmanned systems technology VIII, 62301F

25. Hemami A (1985) Studies on a light weight and flexible robot manipulator. Robotics 1(1):27–36
26. Narang YS, Vlassak JJ et al (2018) Mechanically versatile soft machines through laminar jamming. Adv Func Mater 28(17):1707136

Life Cycle Assessment of Abrasive Flow Machining of 3D Printed Parts: A Comparative Analysis

Muniram Meena, Nitin Dixit, and Varun Sharma

Nomenclature

AFM	Abrasive Flow Machining
FDM	Fused Deposition Modeling
FL	Flood Lubrication
FS	Fumed Silica
MQL	Minimum Quantity Lubrication
PLA	Polylactic Acid
SLM	Selective Laser Melting
XG	Xanthan Gum
SiC	Silicon Carbide

1 Introduction

The surface finish of micron and nano-level is desirable in applications like turbine blades, medical instruments, biomedical implants, automotive, and aerospace industries. The use of conventional methods is exclusive to simple cylindrical and flat surfaces. Therefore, non-conventional processes are preferred over traditional methods of finishing. AFM provides better flexibility to finish components with

M. Meena · V. Sharma
Department of Mechanical and Industrial Engineering, Indian Institute of Technology, Roorkee 247667, India

N. Dixit (✉)
Mechanical Engineering Department, Punjab Engineering College (Deemed to be University), Chandigarh 160012, India
e-mail: ndixit111@gmail.com

V. S. Sharma et al. (eds.), *Machining and Additive Manufacturing*, Lecture Notes in Mechanical Engineering, https://doi.org/10.1007/978-981-99-6094-1_12

complex shapes and inaccessible areas owing to the self-deformable property of abrasive media. The abrasive media accomplishes the finishing action. The constituents of the media are polymer bases, processing oil, and abrasives. The polymer base holds the abrasives during finishing, and processing oil helps to control media flow. Abrasive particles behave as a multipoint cutting tool when these abrasive media extrude through a confined passage between the tooling and workpiece. The properties of abrasive media possess a significant impact on the finishing performance. Polymers, rubber, and hydrogels are used as base polymers to make abrasive media. These bases are needed to be heated sometimes during the media preparation. Different abrasive media have dissimilar constituents and making processes [1–3].

Every raw material, product, waste, production process, and other inputs and outputs involved in the process have some impact on society, the environment, social life, and the economy [4]. Therefore, the manufacturing processes, materials, and products should be sustainable. Sustainability depends upon three dimensions, i.e., environmental life cycle analysis, life cycle costing, and social life cycle assessment [5]. Life cycle analysis (LCA) is a tool to evaluate the ecological impacts and resources used during a product life cycle. The product life cycle involves raw material extraction, manufacturing, and use of application levels, to waste control. The waste control segment includes disposal in addition to recycling. There are three different types of studies in LCA, based on the process involved or product used, cradle-to-grave, cradle-to-gate, and gate-to-gate. The term "product" consists of both goods and services. LCA is an exhaustive appraisal and considers all the elements of the impacts occurring on nature, human health, and environmental resources [6].

In machining, LCA is essential in determining, optimizing, and minimizing the environmental impact of processes, services, and products. LCA has been performed for drilling and milling in Flood Lubrication (FL) and Minimum Quantity Lubrication (MQL) conditions. It was found that the environmental impact was more adverse in FL than in MQL [7]. In a similar line, FL, MQL, and cryogenic cooling have been used to machine Ti-6Al-4 V. MQL was found less hazardous [8]. The ecological and societal assessment of two novel cooling strategies showed that normal minimum quantity cutting fluid condition has a minimum impact on human health and the ecosystem compared to Ranque-Hilsch vortex tube-assisted minimum quantity cutting fluid condition for machining titanium alloy [9].

The literature study reveals that LCA of finishing processes is essential to examine the impact of processes and materials on the environment, society, and economy. Also, no such attempt has been reported regarding the life cycle assessment of the AFM process and its abrasive media. In this research study, LCA has been performed on the AFM process for finishing FDM-printed PLA parts using natural polymer-based and hydrogel-based abrasive media. SimaPro (Version 9.1.0.8) software was used to perform this analysis. The ReCiPe 2016 V 1.04 midpoint(E) and endpoint(E) module was used to evaluate impact assessment.

2 Experimentation

PLA is a biodegradable material used in biomedical applications. Cylindrical PLA parts were printed using an FDM printer available in the lab (Make: AHA 3-D, Model: Proto Centre 999) layer-by-layer. These 3D-printed parts have been finished using an indigenously developed AFM facility [10]. The details of the AFM setup, preparation of abrasive media, LCA inventory, and impact assessment are illustrated in this section.

2.1 Experimental Setup

An indigenously developed AFM setup consists of two hydraulic cylinders, two media cylinders, a power pack assembly, a three-phase power supply, and a control panel. The maximum weight of media used in finishing is 800 gm, based on the volume of the media cylinder.

2.2 Abrasive Media

In the present study, hydrogel-based abrasive media and natural polymer-based abrasive media have been synthesized and utilized in AFM. The processes of preparation of these abrasive media are explained below.

2.2.1 Hydrogel-Based Abrasive Media

In hydrogel-based abrasive media preparation, Xanthan Gum (XG), Fumed Silica (FS), Locust Bean Gum (LBG), Silicon Carbide (SiC) abrasives, and water were used as main constituents. Initially, a solution of XG and water was prepared, and then LBG and FS were introduced as a crosslinker and thickening agent to prepare hydrogel. Lastly, SiC abrasives have been added to prepare abrasive media. A mechanical mixture with a power rating of 300 Watts was used to properly mix media constituents for 45 min. The media preparation consumed 0.075 kWh of electrical energy. The abrasive media can effectively remove material up to 1000 finishing cycles. In this study, each PLA part was finished for 300 cycles. Prepared abrasive media (800 gm) prepared once has been used to finish three parts. Therefore, 266.667 gm abrasive media have been used effectively for each PLA part. The constituent of media to prepare 266.667 gm of abrasives is given in Table 1.

Table 1 Constituents of hydrogel-based abrasive media

Hydrogel-based abrasive media			Natural polymer-based abrasive media		
Constituents	%	Weight (266.667 gm)	Constituents	%	Weight (266.667 gm)
XG	5	13.333	Natural polymer	37.5	100
LBG	2.5	6.667	Dimethyl silicone oil	12.5	33.333
FS	2.5	6.667	Abrasive (SiC)	50	133.333
Water	40	106.667			
Abrasive (SiC)	50	133.333			
Total	100	266.667	Total	100	266.667

2.2.2 Natural Polymer-Based Abrasive Media

Natural polymer, dimethyl silicone oil, and SiC abrasive were used as the main constituents to prepare natural polymer-based abrasive media as tabulated in Table 1. This media was prepared by heating a natural polymer in a hot air oven (power rating 300 watts) to 200°C for 2 h and mixing dimethyl silicone oil and abrasives using a mechanical mixture for 1 h. The preparation of polymer-based abrasive media consumed 1.767 KWh of electrical energy. The same abrasive media has been used to finish three PLA parts.

2.3 Life Cycle Assessment (LCA)

A life cycle assessment tool was used to analyze the environmental emission and the impact of the process on the environment. According to ISO 14040, LCA analysis has four stages: the definition of goal and scope, inventory, impact assessment, and interpretation phases shown in Fig. 1. SimaPro (version 9.1.0.8) software was used for the analysis.

2.3.1 Goal and Scope Definition

This defines the purpose of analysis and system boundary. In the present study, the scope of the work was gate-to-gate, which excluded the transport of raw material, application of the finished product, and waste management of the process and product. The abrasive media, workpiece material, and energy consumption are functional units. The Ecoinvent v1.03 was selected as a database for the current study. Figure 1b shows the scope of the LCA study.

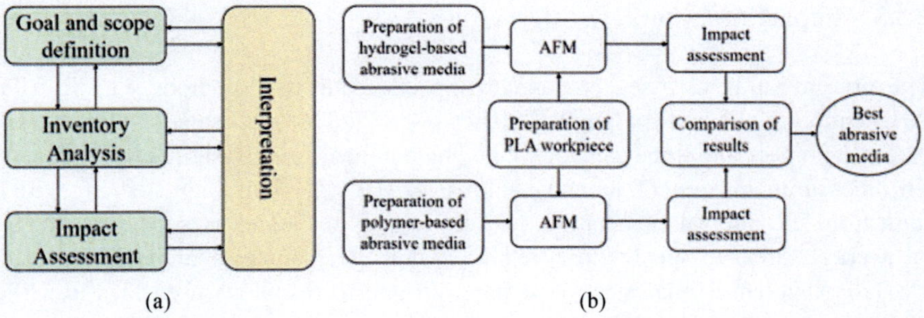

Fig. 1 **a** Stages of LCA. **b**. Scope of the LCA study

2.3.2 Inventory Analysis

The constituents of hydrogel and natural polymer-based abrasive media with quantity are tabulated in Tables 1 and 2, respectively. The inventory flow diagrams of the AFM process using these abrasive media are presented in Fig. 2a and b.

Table 2 Comparison of environmental damage due to AFM of PLA workpiece

Damage category	Unit	Hydrogel abrasive media-based AFM	Natural polymer abrasive media-based AFM
Human health	DALY	1.76E-04	3.65E-04
Ecosystem	Species. yr	1.35E-07	2.80E-07
Resources	USD2013	8.52E-02	0.192

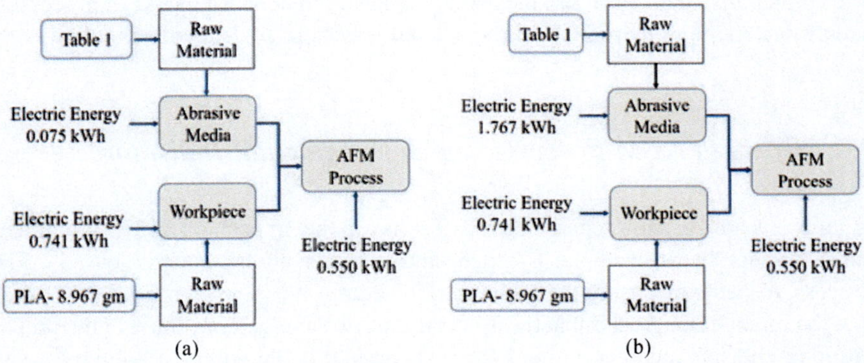

Fig. 2 **a**. Inventory flow diagram of hydrogel media-based AFM. **b**. Inventory flow diagram of natural polymer media-based AFM

2.3.3 Impact Assessment

The present study assessed ecological impacts using two methods, i.e., ReCiPe 2016 endpoint (E) and midpoint(E). The ReCiPe 2016 endpoint(E) method evaluates the impact into three categories, i.e., human health, ecosystem, and resources. The measurement unit of human health is DALY (disability-adjusted life year), indicating the equivalent health loss. The unit of resources is species/year, i.e., an average threat to species in a year. The unit of resources scarcity indicator is USD2013 which estimates the cost incurred due to resources used. The ReCiPe 2016 midpoint(E) method evaluates the impacts into 18 categories. This approach gives a comprehensive analysis involving several categories with different measuring units.

2.3.4 Interpretation

In this phase, the results of the other three steps are interpreted according to the defined goal. After getting all the inventory data related to the processes, introduce it into SimaPro (version 9.1.0.8) software, and get the LCA output. This phase interprets the output of impact assessment into a meaningful result by using world ReCiPe midpoint(E) and endpoint(E) methods and finding the best method that has less impact on the environment.

3 Results and Discussions

The comparative damage assessment analysis has been done for the AFM processes, and evaluated their environmental impact and emissions by LCA are as follows.

3.1 Effect of AFM Processes on Environmental Indicators

Figures 3a and b show characterized results of the impact on the environment category-wise because of the hydrogel and polymer abrasive media-based AFM process, respectively, using the midpoint(E) method. It can be seen from the figures that the major ecological impact was because of two energy consumed in the preparation of abrasive media, printing PLA parts on FDM, and energy used in the AFM process. These results revealed that energy consumption significantly affects the environment.

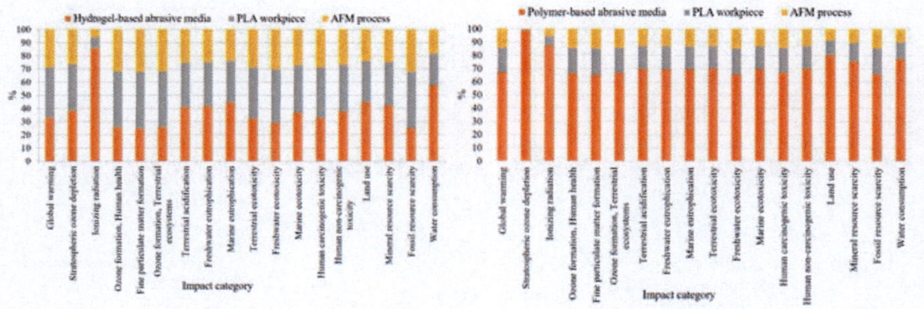

(a) (b)

Fig. 3 **a** Impact due to hydrogel abrasive media-based AFM. **b** Impact due to polymer abrasive media-based AFM

3.2 *Comparison of Environmental Emissions*

Figure 4 compares the total emission of hydrogel media-based AFM and polymer media-based AFM. The major emission categories are the 1,4-DCB, CO2-eq, OIL-eq, and CO-60-Eq. 1,4-DCB and are highest for polymer abrasive media-based AFM. All other emissions are also higher for the polymer abrasive media-based AFM process.

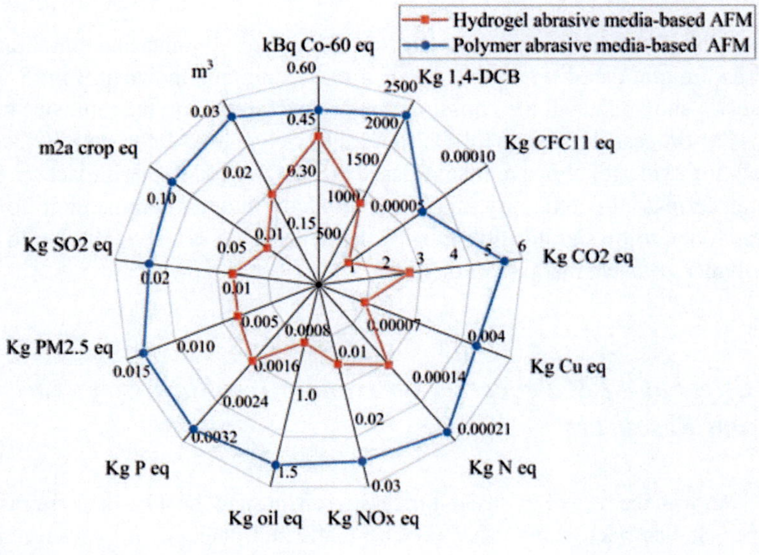

Fig. 4 Comparison of environmental emissions

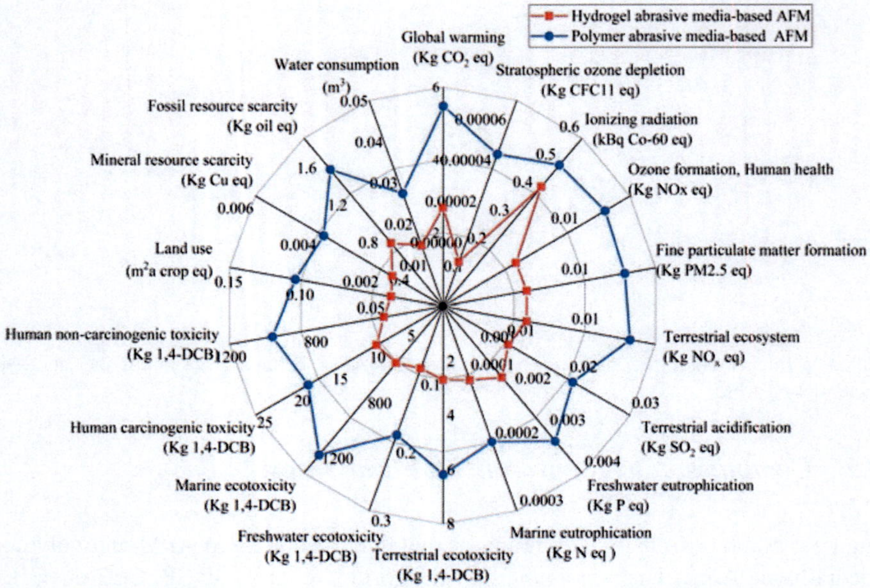

Fig. 5 Comparison of the different environmental indicators

3.3 *Comparison of Environmental Indicators*

Emissions of all the environmental indicators from the midpoint(E) method for both processes are compared with the help of a radar diagram shown in Fig. 5. It has been clearly shown that all the emissions are higher for the polymer abrasive media-based AFM process. CO2 eq., 1,4-DCB eq., NOx eq., CFC11 eq., and SO2 eq. are very less for hydrogel abrasive media-based AFM. Therefore, the impact on global warming, terrestrial ecotoxicity, marine ecotoxicity, human carcinogenic toxicity, etc., was found to be significantly less for the hydrogel abrasive media-based AFM than polymer abrasive media-based AFM.

3.4 *Effect of AFM Processes on Human Health, Ecosystems, and Resources*

Table 2 shows the effect of used processes on human health, ecosystems, and resources. It has been shown that the most affected category is resource. It has a 55.62% higher ecological impact because of polymer abrasive media-based AFM. Also, human health and the ecosystem also have a greater impact because of the polymer abrasive media-based AFM process. Electricity consumption for media preparation is the major contributor to this trend, and it is 95.74% less for hydrogel-based media.

4 Conclusions

The conclusions of the present study include:

- XG abrasive media-based AFM has less environmental damage, i.e. human health, ecosystem, and resources, and very few emissions like CO_2 eq., 1,4-DCB eq., NO_x eq., CFC11 eq., SO_2 eq., etc., than polymer abrasive media-based AFM.
- In preparing hydrogel-based abrasive media, energy consumption is 95.74% less than polymer-based abrasive media.
- Hydrogel abrasive media-based AFM has 51.78% less impact on human health, 51.78% less on the ecosystem, and 55.62% less on resources than polymer abrasive media-based AFM.
- Hence, hydrogel abrasive media-based AFM has excellent potential to reduce environmental impact and is recommended over polymer abrasive media-based AFM.

References

1. Dixit N, Sharma V, Kumar P (2021) Research trends in abrasive flow machining: a systematic review. J Manuf Process 64:1434–1461
2. Dixit N, Sharma V, Kumar P (2021) Development and characterization of xanthan gum-based abrasive media and performance analysis using abrasive flow machining. J Manuf Process 67:101–115
3. Jain VK, Adsul SG (2000) Experimental investigations into abrasive flow machining (AFM). Int J Mach Tools Manuf 40:1003–1021
4. Singh AK, Sharma V (2022) A comparative appraisal of sustainable strategy in ultrasonic assisted grinding of nimonic 80A using novel green atomized cutting fluid. Sustain Mater Technol 32:e00423
5. Tarne P, Traverso M, Finkbeiner M (2017) Review of life cycle sustainability assessment and potential for its adoption at an automotive company. Sustain 9(4):670
6. Singh R, Sharma V (2022) Experimental investigations into sustainable machining of Hastelloy C-276 under different lubricating strategies. J Manuf Process 75:138–153
7. Campitelli A, Cristóbal J, Fischer J, Becker B, Schebek L (2019) Resource efficiency analysis of lubricating strategies for machining processes using life cycle assessment methodology. J Clean Prod 222:464–475
8. Khanna N, Shah P, de Lacalle LNL, Rodríguez A, Pereira O (2021) In pursuit of sustainable cutting fluid strategy for machining Ti-6Al-4V using life cycle analysis. Sustain Mater Technology 29:e00301
9. Gupta MK, Song Q, Liu Z, Pruncu CI, Mia M, Singh G (2020) Machining characteristics-based life cycle assessment in eco-benign turning of pure titanium alloy. J Clean Prod 251:119598
10. Dixit N, Sharma V, Kumar P (2021) Experimental investigations into abrasive flow machining (AFM) of 3D printed ABS and PLA parts. Rapid Prototyp J 28:161–174

Modelling and Simulation of Wire DED Additive Manufacturing Process

Akshay Arjun Pradhan, Rajeev Srivastava, and Abhinav Sarma

Nomenclature

AM	Additive Manufacturing
FEA	Finite Element Analysis
WAAM	Wire Arc Additive Manufacturing
DED	Direct Energy Deposition

1 Introduction

Additive manufacturing methods are classified on the basis of three components viz., the heating power source used (e.g., laser, electron beam, electric arc), the filler material status (powder, wire), and the shape-forming method (selective melting, direction deposition). Due to its inherent cheap cost, high material deposition rate, and direct full-density build-up nature, wire-arc additive manufacturing (WAAM), has been regarded as the most promising AM method of low-cost mass production application for medium- to large-sized components.

Wire Arc Additive Manufacturing (WAAM) constitutes the category of Direct Energy Deposition (DED), where filler wire is constantly melted using a heat source emanating from the nozzle. Metal layers made up of the drops are deposited one on top of the other. This is very similar to how traditional welding is done. The classification of direct energy deposition is shown in Fig. 1.

Although the WAAM is a technique that mass production firms find highly appealing, this process is still in its early stages of development. The large residual

A. A. Pradhan (✉) · R. Srivastava · A. Sarma
Department of Mechanical Engineering, Motilal Nehru National Institute of Technology Allahabad, Prayagraj 211004, India
e-mail: akahay424@gmail.com

V. S. Sharma et al. (eds.), *Machining and Additive Manufacturing*, Lecture Notes in Mechanical Engineering, https://doi.org/10.1007/978-981-99-6094-1_13

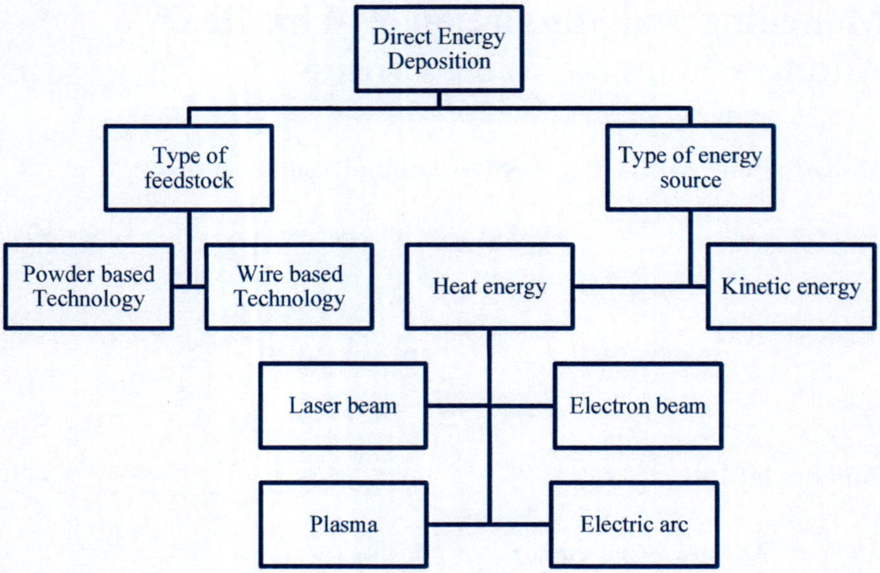

Fig. 1 Classification direct energy deposition

stress, which would result in part distortion, deposition flaws, and build-up failure, is one of the most worrisome problems due to the high heat input during layer deposition. As it is well-known, the temperature gradient produced by the arc in arc deposition mostly induces residual stress.

The present study has developed a finite element model that predicts the temperature fields and residual stress distribution in SUS 304 plates.

2 Methodology

Simulation plays a very crucial role in additive manufacturing processes. We get to know the stresses developing beforehand and gives us an idea about the model that is to be manufactured. Firstly, the 3D model of the plate will be modeled in the ABAQUS. Further, meshing and initial boundary conditions will be applied, as can be seen from the flow chart (Fig. 2) given. The material template is created as per the properties of SUS304 [1].

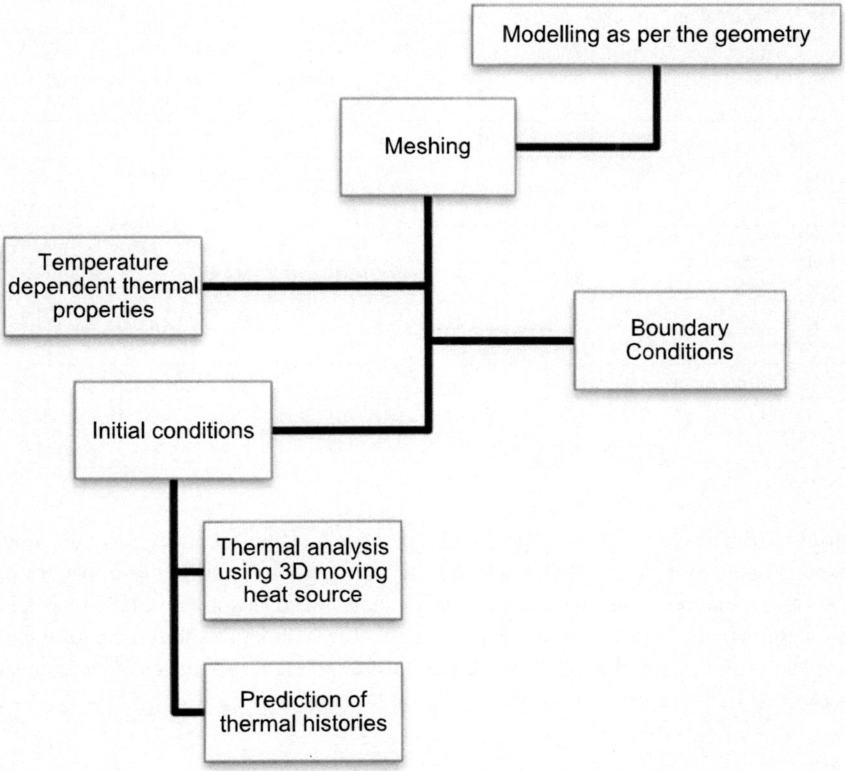

Fig. 2 Flow chart of methodology

3 Numerical Simulation

We can investigate "what if" scenarios and questions through simulation rather than by performing actual system experiments. Before you start printing, it will assist to optimize the setup of your machine, parts, and material combinations, reduce, and perhaps, even eliminate the physical process of trial-and-error testing as a result. It aids in locating obstructions in the movement of materials, information, and goods. Understanding which factors are most crucial to system performance is helpful. Table 1 shows the thermal and mechanical properties of SUS304 material.

3.1 Modeling

The dimensions of the plate used are 30 × 2 × 100 mm. The material used in this simulation is Steel SUS304. The approximate global size of the mesh is 2 mm. The

Table 1 Thermal and mechanical properties of SUS304 [2]

S.no.	Temperature (°C)	Density (Kg/m^3)	Specific heat (J/kg °C)	Conductivity (W/m °C)	Thermal expansion (X10^{-5} °C^{-1})	Yield stress (MPa)	Young's modulus (GPa)	Poisson's ratio
1	0	7900	462	14.6	1.70	265	198.50	0.294
2	100	7880	496	15.1	1.74	218	193	0.295
3	200	7830	512	16.1	1.80	186	185	0.301
4	300	7790	525	17.9	1.86	170	176	0.310
5	400	7750	540	18.0	1.91	155	167	0.318
6	600	7660	577	20.8	1.96	149	159	0.326
7	800	7560	604	23.9	2.02	91	151	0.333
8	1200	7370	676	32.2	2.07	25	60	0.339
9	1300	7320	692	33.7	2.11	21	20.00	0.342
10	1500	7320	700	120	2.16	10	10	0.388

initial conditions of the plate are at 25 °C. The plate has fixed support along the length on both sides. Since the weld bead and material type won't have a big impact on the outcome, the aim of this study is to simply detect variations in residual stresses and distortion. Hence the study doesn't need to go into great depth about the geometry of the bead or the material used. The temperature-dependent properties of the material mentioned in the table are considered during the analysis.

3.2 *Heat Source Model*

In order to simulate a WAAM process accurately, the intricate physical processes that produce heat must be replaced with a mathematical simulation of a heat source with the correct input power distribution. As with all other comparable procedures, the heat source is used to create the weld pool by melting the metal and using the heat generated. In this study, Goldak's double ellipsoidal model (Fig. 3) is used, given in Eq. 1 [3]

$$q_r(x, y, z, t) = \frac{6\sqrt{3Q f_r}}{abC_r\pi\sqrt{\pi}} \exp\frac{-3x^2}{a^2} \exp\frac{-3y^2}{b^2} \exp\frac{-3[z + v(\tau - t)]^2}{C_r^2} \quad (1)$$

where V = arc voltage; $Q = \eta$; VI = power of the arc; η = arc efficiency of the TIG welding process; I = welding current. The fractions f_r and f_f are two different parameters for heat deposition in the front and rear of two ellipsoids. The sum of f_f and f_r is considered as 2 ($f_f = 1.4, f_r = 0.6$) [1, 6].

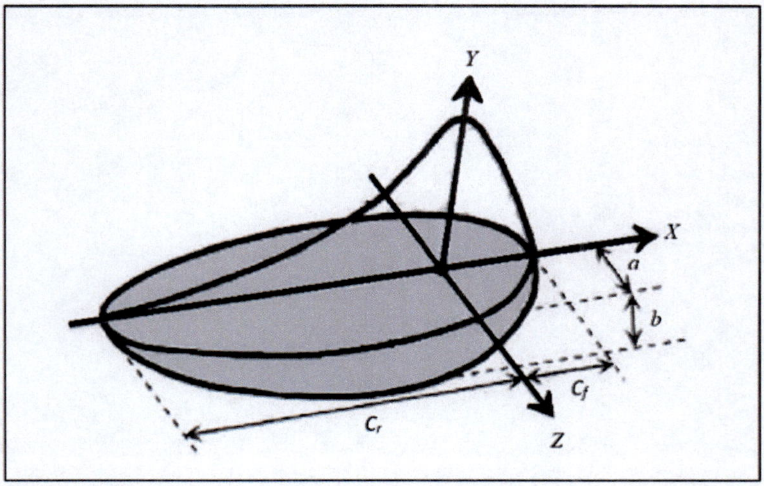

Fig. 3 Goldak moving heat source model [1]

3.3 Simulation

The Python run script was developed to run the moving heat source model. All the parameters of the model discussed above are given with the help of the Python run script. It includes dimensions of the heat source model, values for calculating the heat flux, and Goldak's heat source model equation.

The simulation shows the Von Mises stress, residual stress developing throughout the moving heat source. The maximum yield stress developed in the plate is 679 MPa and the maximum principle stress is 285 MPa. The residual stresses developing in the plate are less than the yield stress of the material. As time increases, the peak temperature values the decreased at the weld zone. These peak temperature values (Fig. 5) and accumulation of more heat can have effects on developing more stress near the molten pool zone [4]. The graphs are plotted at 50 s time interval and along the moving heat source model, i.e., in the longitudinal direction. The path distance is 100 mm. The Yield stress developing at 50 s time interval is 294 MPa and residual stress is 153 MPa. Further, residual stress is maximum at 900mm approximately from the starting of the heat source model [5].

The above graphs (Fig. 6) are plotted at 50 s time intervals and along the moving heat source model, i.e., in the longitudinal direction. The path distance is 100 mm. The yield stress developing at a 50 s time interval is 294 MPa and the residual stress is 153 MPa can be seen from Fig. 4a and b. Further, residual stress is maximum at 90 mm approximately from the start of the heat source model.

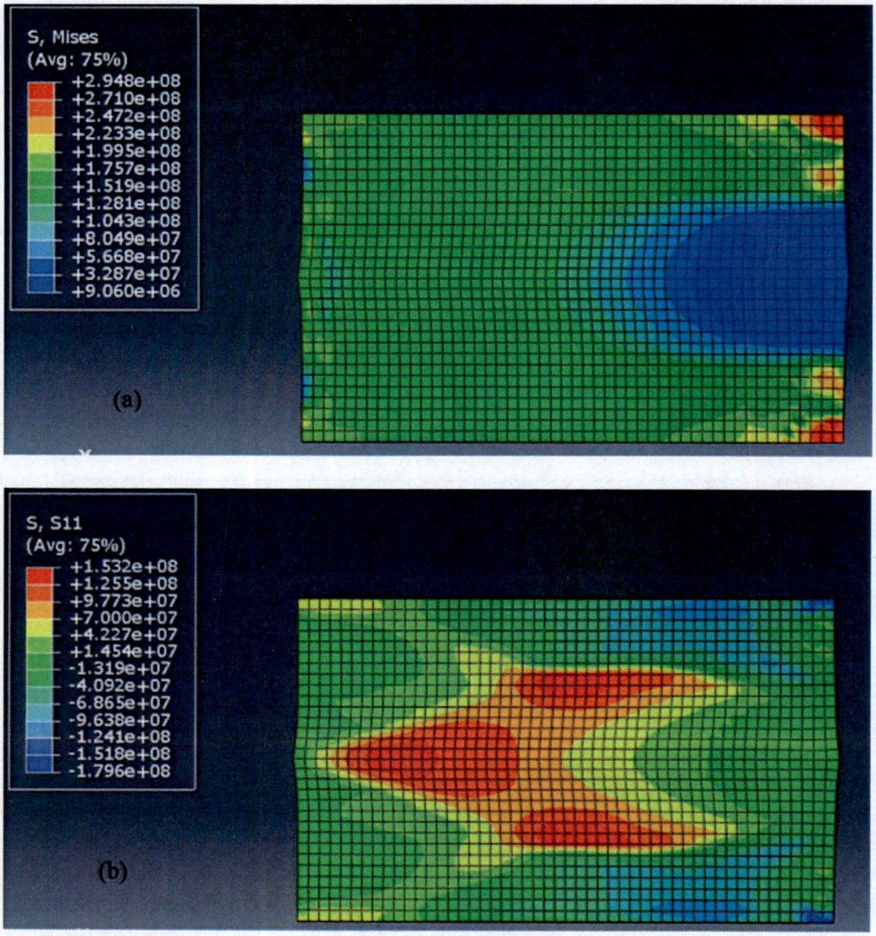

Fig. 4 Goldak's moving heat source model **a** Von Mises Stress **b** Residual stress

4 Conclusions

The present study has focused on how well the ABAQUS CAE 2020 AM can handle rigors of a two-plate geometry full process simulation. Goldak's heat source model is used to predict the residual stresses and thermal cycles on the SUS 304 and the Python run script is developed to run the model. Initially, the plate is modeled as per the dimensions 30 × 2 × 100 mm with a material SUS 304 and then after, a partition is created to convert it into two plates. The residual stress developing in the plate during the moving heat source is about 381 MPa. The maximum yield stress developed in the plate is 679 MPa and the maximum principle stress is 285 MPa. The study builds that the residual stresses in the deposited material are accurately predicted using Goldak's 3D heat source model.

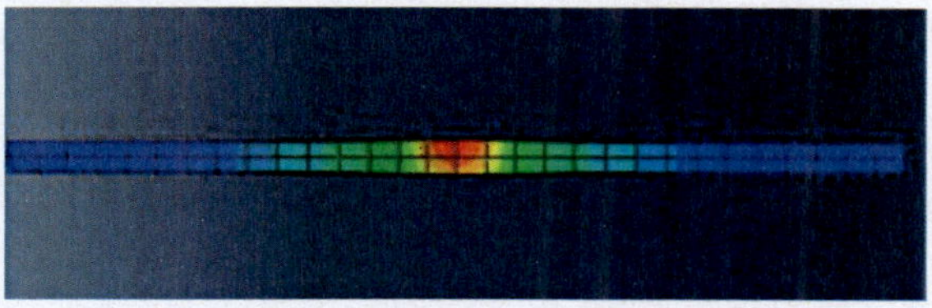

Colour Indicates temperature distribution	Description
	Maximum Temperature
	Molten pool temperature
	Heat affected zone
	Less affected or unaffected zone

Fig. 5 Color indication for temperature distribution

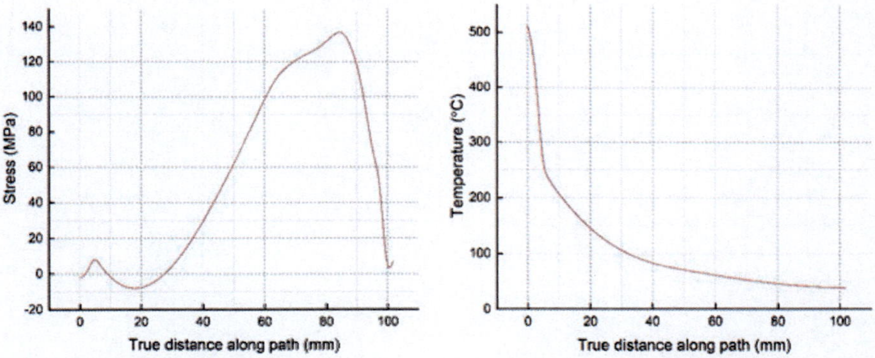

(a) Residual stress distribution along the length.

(b) Temperature distribution along the length.

Fig. 6 **a** Residual stress distribution along the length. **b** Temperature distribution along the length

References

1. Balram Y, Sridhar Babu B, Vishnu Vardhan T et al (2019) Residual stress analysis of dissimilar tungsten inert gas weldments of AISI 304 and Monel 400 by numerical simulation and experimentation. Mater Today: Proc. https://doi.org/10.1016/j.matpr.2019.07.639
2. Deng D, Murakawa H (2006) Numerical simulation of temperature field and residual stress in multi-pass welds in stainless steel pipe and comparison with experimental measurements. Comput Mater Sci 37(3):269–277. https://doi.org/10.1016/j.commatsci.2005.07.007
3. Amal MS, Justus Panicker CT, Senthilkumar V (2022) Simulation of wire arc additive manufacturing to find out the optimal path planning strategy. Mater Today: Proc 66:2405–2410. https://doi.org/10.1016/j.matpr.2022.06.338
4. Colegrove P, Ganguly S (2017) Residual stress characterization and control in the additive manufacture of large scale metal structures. Residual Stress 2016:455. https://doi.org/10.21741/9781945291173-77.

5. Venkatkumar D, Ravindran D (2016) 3D finite element simulation of temperature distribution, residual stress and distortion on 304 stainless steel plates using GTA welding. J Mech Sci Technol 30(1):67–76. https://doi.org/10.1007/s12206-015-1208-5
6. Goldak J, Chakravarti A, Bibby M (1984) A new finite element model for welding heat sources. Metall Trans B 15(2):299–305. https://doi.org/10.1007/bf02667333

An Overview of the Recent Advances in Additive Manufacturing, and Its Scope for the Benefit of Research and Development in Commercial Vehicle Industries

P. Sriram Madhav, R. Revathi, and D. Keerthi Vasan

1 Introduction

Industry 4.0 is the latest prevailing Industrial Revolution, and it differs from earlier ones in that it was based on advances in energy innovation and development, whereas Industry 4.0 is based on a combination of cutting-edge technologies that helped to advance the manufacturing sector into this new era. These cutting-edge technologies include, among others, additive manufacturing (AM), robots, big data, and the Internet of Things. It is not possible to exaggerate the importance of AM technologies in Industry 4.0. It is the primary driver of Industry 4.0, and other cutting-edge technologies are providing crucial assistance for the effective functioning of AM [1].

It is evolving as a technology that will enable and ensure a wide range of new applications. Primarily, the material availability, production speed, and resolution of 3D rapid prototyping techniques must be taken into account for any given application. This overview covers the principles of 3D printing methods as well as the current development of unique 3D printing materials. It should be emphasized that the diversity of 3D printing materials arises from the wide range of 3D printing systems available, and no new printers or methods for unique materials have gone beyond the seven categories outlined in the ISO/ASTM standard.

3D printing, on the other hand, should never be viewed as a stand-alone process; rather, it is becoming an intrinsic element of a multiprocess system or an integrated process of numerous systems to match the development of innovative materials and new product needs. Standard 3D printing techniques that are the most widely used

P. Sriram Madhav (✉) · R. Revathi · D. Keerthi Vasan
Department of Production Engineering, PSG College of Technology, Coimbatore 641004, India
e-mail: srimadp1123@gmail.com

R. Revathi
Department of Product Engineering, Daimler India Commercial Vehicles, Chennai 603109, India

V. S. Sharma et al. (eds.), *Machining and Additive Manufacturing*, Lecture Notes in Mechanical Engineering, https://doi.org/10.1007/978-981-99-6094-1_14

include vat photopolymerization (stereo lithography), powder bed fusion, material extrusion; PLT (paper lamination technology), DED (directed energy deposition), poly jet printing, and holography-infused sintering.

These techniques all build parts layer by layer and provide various options for price, feature detail, and materials. Since there is no requirement for a mold/die, and design concepts are developed into products through digital prototyping, 3D printing is best suited for the mass customization of complex geometric shapes. This is in contrast to conventional manufacturing technologies, such as casting, forging, and machining, which are best suited for the mass production of identical commodity items. In addition to this, the layer-by-layer approach makes it possible to combine several components into a single piece, eliminating the need for additional assembly steps.

2 Am in Commercial Vehicle Industries

2.1 Overview

Recent patent expirations for the initial 3D printing technologies have led to a surge of low-cost desktop systems that make 3D printing more accessible to the public. The manufacturing sector has realized the benefits of these advancements and is investing in production systems to produce complicated parts for jet engines, specially customized vehicle bodywork, and even medications. Despite the fact that standard 3D printing technology has advanced to the point where it is now possible to print in a variety of materials [1, 2].

One sector with an interest in more extensive AM uses is the commercial vehicle sector. Investment groups are the typical purchasers of commercial cars, and these organizations must tailor the vehicles to meet their particular investment needs. In addition, the infrastructure and legal requirements for various markets vary. As a result, in order to meet these demands, manufacturers and suppliers must broaden their range of products and create more customized, non-standard parts in small batches. In order to handle the variety of products and the manufacture of small batches, they must also set up more systems and procedures, which eventually increases the complexity of their manufacturing processes. AM has great promise for assisting commercial vehicle manufacturers to increase their level of product and system flexibility [2].

For instance, on the level of the individual product, AM enables the development and production of geometrically complex features by using topology optimization and light weighting, and as a result, manufacturers can design and produce more complex parts to enhance the overall performance of their vehicles. Another significant advantage of using additive manufacturing for tooling is the ability for manufacturers to build assembling tools in one-off or small-batch production at a cheaper cost. AM allows cost reductions and a shortened process chain at the system level.

Since more equipment and production procedures are needed in conventional manufacturing, items with more geometrical complexity have higher manufacturing costs and longer process chains [3].

Companies have been using 3D printers to make prototypes and concept models for more than 30 years, largely for "fit and form" inspections. Because we were aware that plastic 3D-printed pieces would not be as durable as the finished injection-molded parts, functionality was not a priority. Companies may now additively produce functioning prototypes instead of merely fit-and-form models since we can now print metallic components.

2.2 *Material Availability—Metals*

List of materials available for metal additive manufacturing applicable for commercial vehicle industries is listed (Table 1). The properties of various materials listed in Table 1 are shown in Fig. 1. ASTM F2792-12a is Standard Terminology for Additive Manufacturing Technologies.

Compared to polymers, metal additive manufacturing plays a much more important role in commercial vehicle industries. This is because, majority of the components require high strength and hardness, and plastics cannot attain those parameters. On the other hand, even metal powders are very limited as depicted above; moreover, the setup cost is very high compared to the conventional manufacturing techniques. The above alloys are named as per their primary or base metal, with an indication of other elements in order of mass percent [9].

Table 1 AM metal powder list

S. no.	Alloy	Description
1	AlSi10Mg	Hardenable aluminum-based alloy
2	18Ni300	Maraging steel
3	13Ni400	Maraging steel
4	SS316L	Stainless steel (with low carbon content)
5	17-4PH	Martensitic stainless steel
6	IN625	Inconel with higher level of chromium and molybdenum than grade 718
7	IN718	Inconel
8	CoCrMo	Cobalt chromium alloy
9	Ti64AlV	Titanium alloy

Properties \ Materials	AlSi10Mg	18Ni300	13Ni400	SS316L	17-4PH	IN718	IN625	CoCrMo	Ti6Al4V
Hardness*	119 ± 5 HB	53 HRC	56-60 HRC	163 HB	40 HRC	47 HRC	35 HRC	45 HRC	36 HRC
Density	2.67 g/cm^3	8.1 g/cm^3	8.08 g/cm^3	7.9 g/cm^3	7.75 g/cm^3	8.15 g/cm^3	8.44 g/cm^3	8.4 g/cm^3	4.43 g/cm^3
Melting range	570-650 °C	530-1415 °C	530-1415 °C	1400-1450 °C	1404–1440 °C	1290-1350 °C	1370-1430 °C	1100-1330 °C	325-400 °C
Elastic modulus	180 ± 20 GPa	185 ± 7 GPa	179 ± 8 GPa	180-185 GPa	190-193 GPa	205 GPa	206 GPa	123 GPa	114 GPa
Heat treatment	At 300 °C for 2 hours, and air quenching	At 490°C for 6-10 hours, depending on the part size	Precipitation hardening at 525-540 °C for 3-6 hours	Stress relief at 650 °C, solution treatment at 1100 °C	Hardening at 593 °C for 4 hours, air quenching	Solution annealing at 990 °C, ppt. hardening at 620 °C	Stress relief at 595 °C, solution annealing at 1095 °C	Heat-treated at 1200 °C for 2 hrs. combined with water cooling	Solution treatment between 425-650°C for 2 hrs.
Perks	For complex geometries, high corrosion resistance and strength	Superior strength and toughness without losing ductility	Good machinability with high corrosion resistance	Low proportion of carbon imparts even better corrosion resistance	Martensitic steel with high strength and hardness	Greater strength and hardness, less corrosion resistance	Higher levels of Cr and Mo makes it more corrosion resistant	Higher resistance to wear and fatigue	High strength and corrosion resistance, can withstand high temperatures

Fig. 1 Metal additive manufacturing material data sheet [4–8]

2.3 *Material Availability—Plastics*

Some of the most commonly used polymer materials for additive manufacturing include:

i. **Polypropylene (PP)**: This group of materials, which makes up around 20% of the world's output of plastic components, is the second-most significant form of plastic. E.g., Wiper cover, bezel, INS dashboard, and mirror mounting bracket.
ii. **Acrylonitrile Butadiene Styrene (ABS)**: ABS is a widely used thermoplastic material known for its strength, durability, and heat resistance. E.g., Styling parts and aero corner.

These are just a few examples of the polymer materials that are commonly used in additive manufacturing. There are many other materials available, each with their unique properties and applications. The choice of material depends on the specific requirements of the part being produced and the printing technology being used.

2.4 *Comparing Polymer and Metal Additive Manufacturing*

Polymer and metal additive manufacturing (AM) are two distinct approaches to 3D printing that use different materials and printing technologies. There are several key differences between polymer and metal AM that affect the production process, the properties of the finished parts, and the applications in which they are used.

i. **Printing technology**: Polymer AM typically uses extrusion-based printing technology, where molten polymer material is extruded through a nozzle to create the part layer by layer. Metal AM, on the other hand, typically uses a powder bed fusion technique, where a laser or electron beam is used to selectively melt and fuse metal powder to build the part layer by layer.

ii. **Material properties**: Polymer materials typically have lower strength and stiffness compared to metals, but they offer advantages such as low density, flexibility, and ease of processing. Polymer parts are often used for non-structural components, such as housings, enclosures, and prototypes. Metal materials, on the other hand, offer high strength, stiffness, and thermal conductivity, and are commonly used for functional and structural parts in industries such as aerospace, automotive, and medical.

iii. **Post-processing**: Polymer parts may require less post-processing compared to metal parts, as polymer parts typically have a smoother surface finish and may not require additional heat treatment or polishing. Metal parts may require additional post-processing steps such as heat treatment, machining, or surface finishing to achieve the desired mechanical properties and surface finish.

iv. **Cost**: Polymer AM is typically less expensive compared to metal AM, as the materials and equipment used in polymer AM are generally less expensive. However, the cost of metal AM is decreasing as the technology improves and becomes more widely adopted.

Overall, both polymer and metal additive manufacturing offer unique benefits and limitations. The choice of polymer or metal additive manufacturing in the automotive industry depends on the specific application and the performance requirements of the part being produced [10]. Both technologies have unique advantages and limitations, and they can be used in combination to produce high-quality, cost-effective automotive components [4].

3 Am in Research and Development

With additive manufacturing, researchers can create prototypes much more quickly and with fewer resources than traditional manufacturing methods. This allows for faster iterations and testing, ultimately leading to faster development times and more rapid innovation. It enables researchers to create highly customized designs that are not possible with traditional manufacturing methods. This can be useful in developing specialized components or medical devices tailored to an individual's unique anatomy. AM allows for complex shapes and internal structures to be produced that would be difficult or impossible to achieve using traditional manufacturing methods. This flexibility in design can lead to improved functionality and performance.

AM can significantly reduce waste in the research and development process by using only the exact amount of material needed to create a part or component. This is particularly important when working with expensive or rare materials. AM can often be more cost-effective than traditional manufacturing methods for small production runs or for the creation of prototypes. This can be useful in the early stages of research and development when budgets are often limited. It allows quick and easy changes to be made to a design or prototype. This means that researchers can iterate on a design more easily and quickly than with traditional manufacturing methods, leading to a more refined final product. AM has enabled the development of new materials and structures, leading to advancements in fields such as aerospace, biomedical engineering, and materials science.

4 Research Gap

Additive manufacturing has the potential to revolutionize the automotive industry by enabling the production of complex parts with less waste and shorter lead times. However, despite the numerous benefits of additive manufacturing, there are still several research gaps that need to be addressed to fully realize its potential in the automotive industry. One of the major research gaps is the need to develop new materials that are suitable for use in additive manufacturing processes. Most of the materials used in the automotive industry, such as steel, aluminum, and carbon fiber, are difficult to print using conventional additive manufacturing techniques. Therefore, there is a need to develop new materials that can be used in 3D printing and meet the specific requirements of the automotive industry, such as high strength, durability, and temperature resistance.

Another research gap is the need to optimize the design and production process of additive manufacturing in the automotive industry. While additive manufacturing offers the potential for significant improvements in the production process, it also requires a new approach to design and production. Currently, there is a lack of standardized design and production processes for additive manufacturing, which can lead to increased costs and longer lead times.

Furthermore, there is a need to address the issue of scalability in additive manufacturing.

While 3D printing is ideal for producing small batches of parts or prototypes, it is still challenging to scale up the process for mass production. This is due to the limitations of current additive manufacturing technology, such as slower printing speeds and the limited size of the printing bed. Finally, there is a need for more research on the economic feasibility of additive manufacturing in the automotive industry.

While there are many potential benefits of additive manufacturing, such as reduced material waste, lower tooling costs, and faster production times, the initial investment costs for additive manufacturing equipment and materials can be high. Therefore, there is a need for further research to evaluate the economic feasibility of additive manufacturing in the automotive industry and to develop business models that can effectively incorporate this technology.

5 Conclusions

Additive manufacturing is increasingly being used in the commercial vehicle industry for a variety of applications. One of the most significant benefits of additive manufacturing is that it allows for the production of complex parts with less waste and shorter lead times, which can be particularly advantageous for the commercial vehicle industry. One application of additive manufacturing in the commercial vehicle industry is the production of parts for prototyping and testing. 3D printing allows engineers and designers to quickly create prototypes of new commercial vehicle parts and test them for fit, form, and function. This can help to accelerate the design and development process and reduce the time and cost associated with traditional prototyping methods.

Another application of additive manufacturing in the commercial vehicle industry is the production of replacement parts. Commercial vehicles often operate in harsh environments, which can lead to wear and tear on parts over time. With 3D printing, replacement parts can be produced on demand, reducing the need for a large inventory and decreasing the downtime associated with waiting for replacement parts to arrive. In addition, additive manufacturing can be used to produce lightweight components for commercial vehicles, which can help to improve fuel efficiency and reduce emissions. 3D printing can be used to produce parts with complex geometries and internal structures that would be difficult or impossible to produce using traditional manufacturing methods. Overall, the use of additive manufacturing in the commercial vehicle industry has the potential to improve the efficiency, reliability, and sustainability of commercial vehicle manufacturing and maintenance. However, there are still several research gaps and technical challenges that need to be addressed to fully realize the potential of additive manufacturing in the commercial vehicle industry.

References

1. Yi L, Gläßner C, Aurich JC (2019) How to integrate additive manufacturing technologies into manufacturing systems successfully: a perspective from the commercial vehicle industry. J Manuf Syst 53:195–211
2. Vasco JC (2021) Additive manufacturing for the automotive industry. In: Additive manufacturing. Elsevier, pp 505–530
3. Leal R, Barreiros FM, Alves L, Romeiro F, Vasco JC, Santos M, Marto C (2017) Additive manufacturing tooling for the automotive industry. Int J Adv Manuf Technol 92:1671–1676
4. Muhammad MS, Kerbache L, Elomri A (2022) Potential of additive manufacturing for upstream automotive supply chains. Supply Chain Forum Int J 23(1):1–19. Taylor & Francis
5. Cooke A, Slotwinski J (2012) Properties of metal powders for additive manufacturing: a review of the state of the art of metal powder property testing. Gaithersburg, MD, USA: US Department of Commerce, National Institute of Standards and Technology
6. Moghimian P, Poirié T, Habibnejad-Korayem M, Zavala JA, Kroeger J, Marion F, Larouche F (2021) Metal powders in additive manufacturing: a review on reusability and recyclability of common titanium, nickel and aluminum alloys. Addit Manuf 43:102017
7. Mooney B, Kourousis KI (2020) A review of factors affecting the mechanical properties of maraging steel 300 fabricated via laser powder bed fusion. Metals 10(9):1273

8. Sulong AB, Ramli MI, Hau SL, Sahari J, Muhamad N, Suherman H (2013) Rheological and mechanical properties of carbon nanotube/Graphite/SS316L/polypropylene nanocomposite for a conductive polymer composite. Compos Part B: Eng 50:54–61
9. Bayley J (1991) Alloy nomenclature. Dress Accessories c, 13–17
10. Richardson M, Haylock B (2012) Designer/maker: the rise of additive manufacturing, domestic-scale production and the possible implications for the automotive industry. Comput-Aided Des Appl PACE 2:33–48

Design and Development of Customized Helmet for Military Personnel

Akhilesh Misra, Rajeev Srivastava, and Abhinav Sarma

Nomenclature

TBI	Traumatic Brain Injury
IED's	Improvised Explosive Devices
PASGT	Personnel Armor System for Ground Troops
ACH	Advanced Combat Helmet
PBO	Poly-Phenylene Benzobis Oxazole

1 Introduction

The helmets are normally manufactured to some standard sizes. There is a need to use customization and smart manufacturing concepts for the soldiers safety and comfort. The mass customization concepts (can be seen in Fig. 1 [1]) can be used in the manufacturing of customized helmets, specifically for soldiers who are serving at the borders of the country and facing issues with their helmets. A certain quantity of individually designed soldier helmets can be manufactured through collaborative customization by understanding their problems and preferences. A helmet that incorporates protection mechanisms (against ballistics), sensors, night vision equipment, etc. is necessary for modern military operations and technology-driven warfare tactics [2]. Many factors are considered during ballistic helmet design, including comfort, weight, and fit [3]. Helmets, after continuous use, become heavy and unstable for many users, which is why some soldiers do not use helmet when necessary [4]. These factors affect the soldier's decision about the use of the helmet.

A. Misra (✉) · R. Srivastava · A. Sarma
Department of Mechanical Engineering, Motilal Nehru National Institute of Technology Allahabad, Prayagraj, UP 211004, India
e-mail: akhilesh.2021cc22@mnnit.ac.in

V. S. Sharma et al. (eds.), *Machining and Additive Manufacturing*, Lecture Notes in Mechanical Engineering, https://doi.org/10.1007/978-981-99-6094-1_15

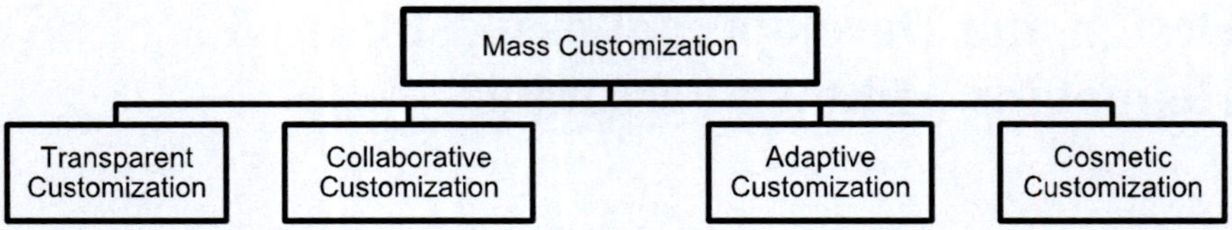

Fig. 1 Types of mass customization [1]

Table 1 Types of blast injuries and their causes [2, 5]

Type of blast injury	Cause of injury
Primary blast injury	The passage of the shock wave through the skull, with the skull absorbing incidental pressure, results in primary blast injuries
Secondary blast injury	Blast winds travel at high speeds during IED (improvised explosive device) explosions and bombing events, and the debris driven by these blast winds results in secondary blast injury
Tertiary blast injuries	The blast winds accelerate the body parts with high intensity. Because the inertia of the skull is greater than that of the brain tissue, high-shearing strains are generated in the intracranial region. This results in tertiary blast injuries
Quaternary blast injuries	After an explosion, high-temperature heat gases are released. This results in burns and respiratory injuries brought on by breathing hazardous gases and fumes. The brain tissues are permanently damaged due to the heating of the skull by these hot gases and fumes

1.1 Traumatic Brain Injury

Traumatic brain injury (TBI), also known as intracranial injury, is damage to the brain caused by external forces, resulting in either temporary or permanent impairment of brain functions. The following causes (can be seen in Table 1) have been identified for blast-induced TBI, based on the existing experimental studies.

1.2 Materials Used in the Manufacture of Military Helmets

The following materials are normally used for the manufacture of military ballistic helmets.

1.2.1 PBO

The Poly-phenylene Benzobis Oxazole (PBO) fibre is commercially known as Zylon®. The tensile modulus of Zylon® fibre is roughly three times that of aramid fibre. PBO fibre's ballistic qualities are substantially degraded by their low resilience to hot and humid environments, which restricts their use in combat helmets [6].

1.2.2 Spectra 1000

Spectra® is a manufactured fibre made from ultra-high molecular weight polyethylene. This remarkably durable material is one of the world's strongest and lightest fibres. The substance has a specific strength that is 40% higher than aramid fibre, 40% more durable than polyester, and ten times stronger than steel. Additionally, it is immune to ultraviolet light (UV) degradation [6].

1.2.3 Dyneema®

The Dyneema® fibre originates from an ultrahigh molecular weight polyethylene solvent spinning process. On a weight-for-weight basis, it is up to 15 times stronger than steel and up to 40% stronger than aramid fibres. Therefore, Dyneema® fibre is an important component in body armor [6].

1.2.4 Kevlar

Kevlar® is a heat-resistant para-aramid synthetic fibre with a molecular structure of many inter-chain bonds that make Kevlar® incredibly strong. On an equal weight basis, Kevlar® has a tensile strength ten times greater than steel. When a bullet or other high-velocity projectile hits Kevlar®, the fibres essentially catch the projectile while absorbing and dissipating its energy[6].

1.3 Types of Helmets

Combat helmets have evolved considerably over the years. There are various types of combat helmets that are used by military personnel, as shown in Table 2.

2 Methodology

The helmets used by the defense units are manufactured in general sizes with standard measurements, which sometimes causes comfort issues (tight or loose fit) for some of the soldiers, due to which they avoid using helmets even when it is necessary. A specific number of helmets can be manufactured for the soldiers facing issues with the size of the helmet by taking measurements of their heads. The 3D model of the

Table 2 Types of ballistic helmets [6]

Helmet type	Visual representation	Material	Features	Weight of the helmet
PASGT [personnel armor system for ground troops]		It is a one-piece composite structure made of Kevlar aramid fibre	It Protects against shrapnel and various handgun bullet impacts. These helmets allow troops to wear a face shield and other tactical equipment along with their helmets	The weight of PASGT helmet ranges from 1.4 to 1.9 kg
ACH [advanced combat helmet]		ACH helmet is made of polyethylene	It has increased blunt impact and ballistic protection, better compatibility with mission equipment, better three-dimensional sound localization, and reduced combat injuries	The weight of ACH helmet is under 1.13 kg
MUKUT advanced combat helmet		It is an advanced combat helmet type Kevlar helmet	It is used by Army, Air force, and Paramilitary units in India. It is effective against 9 mm rounds. It has no holes and no drill technology giving it 360° protection and pads are available to protect against brain injuries	The weight of MUKUT advanced combat helmet is under 1.2 kg
PATKA		PATKA is made of fiberglass	PATKAs were actually been developed for Sikh soldiers, as they always wear their 'Pagdi' (Turban). So, initiation was to make a helmet, which could comfortably fit into the soldiers head	The weight of the PATKA helmet ranges from 1.4 to 1.5 kg

head can be obtained by photogrammetry method, and helmets can be modelled with reference to the 3D head model obtained. In this study, the methodology that was followed in the design and analysis of customized ballistic helmets can be seen in Fig. 2.

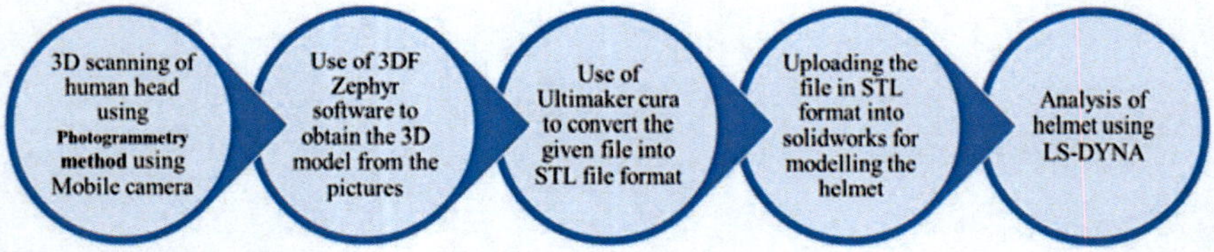

Fig. 2 Methodology chart

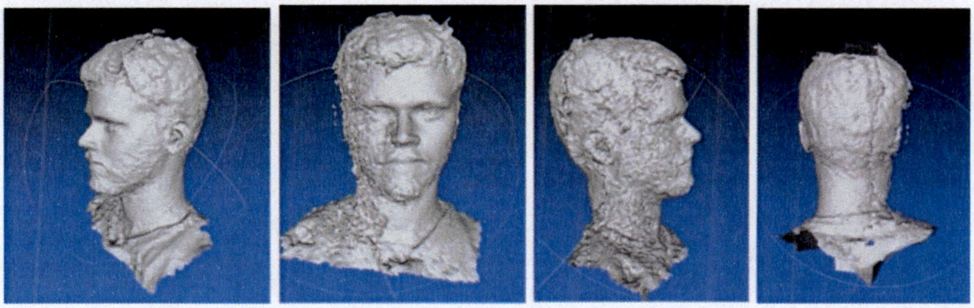

Fig. 3 Human head model obtained from 3DF Zephyr

2.1 3D Scanning of Human Head Using Photogrammetry Method

By using the photogrammetry method, pictures of the human head from different views are taken and uploaded into the software (3DF Zephyr) to obtain the 3D model of the human head. The human head model is obtained, which serves as the foundation for designing a customized helmet.

After uploading the pictures of the human head into 3DF Zephyr software, the human head model is obtained, as shown in Fig. 3.

2.2 Modelling of Helmet

The obtained head model is used as a reference, and a helmet is designed for the human head model using the modelling software (SOLIDWORKS). The modelled helmet is camouflaged to deceive the enemy's viewpoint. The different views of the helmet model can be seen in Fig. 4.

Fig. 4 Different views of helmet model **a** isometric view, **b** cross-sectional view, **c** side view, **d** front view, **e** top view, **f** bottom view

3 Results and Discussions

By using photogrammetry, a model of the human head is generated, and the model is used to design a helmet. The strength of the helmet is evaluated by an impact analysis. In order to conduct the analysis, the 9 $\times$ 19 Parabellum bullet is used as a projectile because it is the most widely used military handgun cartridge in the world [7]. The helmet shell is considered to be made of Kevlar material, and the Kevlar material properties are taken into account based on Chang-Chang failure criteria [7]. Ansys LS-DYNA has been used to conduct the analysis and obtain results.

Figure 6 shows the deflection of the helmet when the bullet strikes it. The maximum deflection when the bullet strikes the helmet from the side is 51.68 mm, and the maximum deflection when the bullet strikes the helmet from the top is 32.23 mm. Various studies have been conducted on ballistic helmets so far, which show that the maximum deflection when the bullet strikes the helmet from the side is 27.4 mm [8]. From this, it can be inferred that the ballistic helmet designed in this work meets the strength criteria.

Figures 7 and 8 show the energy conversion when the bullet strikes the helmet. Initially, when time $t = 0$, the total energy is equal to the kinetic energy of the bullet. However, when the bullet strikes the helmet at time $t = 0.05$ s, the kinetic energy decreases as some of the kinetic energy is absorbed by the helmet, while the internal energy of the system increases as the helmet absorbs the most amount of internal energy and the bullet also absorbs a huge amount of internal energy prior to the helmet. The total energy of the system is simply the sum of its internal, kinetic, and potential energies. In this case, the total energy of the system is decreased as the potential energy is absent and the decrease in kinetic energy is greater than the increase in internal energy.

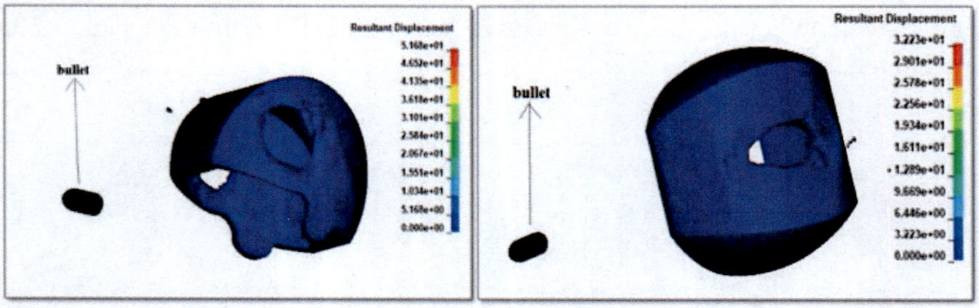

Fig. 6 Resultant displacement when bullet strikes the helmet from side and top

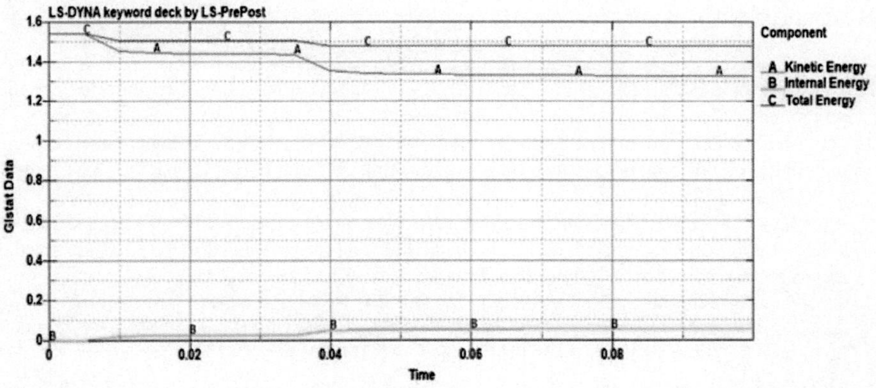

Fig. 7 Kinetic energy, internal energy and total energy plots when the bullet strikes the helmet from sideway

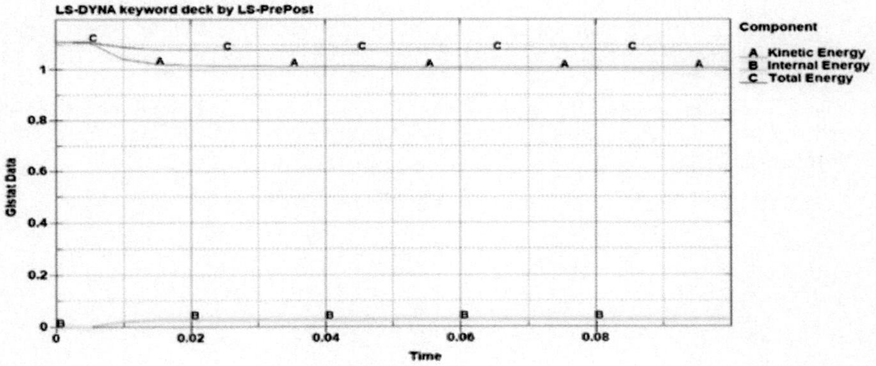

Fig. 8 Kinetic energy, internal energy and total energy plots when bullet strikes helmet from the top

4 Conclusions

In this study, a human head model was generated using the photogrammetry method, and from that model, a helmet was designed and an analysis was performed on the helmet. It was found that the deflection of the helmet when the bullet strikes it from the side ranges from 0 mm to 51.68 mm, and the deflection when the bullet strikes from the top ranges from 0 mm to 32.23 mm, which means that the strength of the helmet at the sides is greater than the top portion. The ballistic helmet meets the strength criteria. In a similar manner, various head scans can be taken by the photogrammetry method, and a wide range of data related to human head measurements can be obtained. Head models obtained can be used in the design and manufacture of helmets based on the dimensions of the head. Customized helmets for military personnel can be manufactured in this manner.

References

1. Da Silveira G, Borenstein D, Fogliatto HS (2001) Mass customization: literature review and research directions. Int J Prod Econ 72(49)
2. Kulkarni SG, Gao X, Horner SE, Zheng JQ, David NV (2013) Ballistic helmets—their design, materials, and performance against traumatic brain injury. Compos Struct 101:313–331. https://doi.org/10.1016/j.compstruct.2013.02.014
3. Ivins GBJ et al (2007) How satisfied are soldiers with their ballistic helmets? A comparison of soldiers' opinions about the advanced combat helmet and the personal armor system for ground troops helmet. Mil Med 172(June):586–591
4. Catapan MF, Okimoto MLLR, Santana FE, Silva MA, Rodrigues YW (2015) Anthropometric analysis of human head for designing ballistic helmets. Procedia Manuf 3(Ahfe):5475–5481. https://doi.org/10.1016/j.promfg.2015.07.689
5. Cernak I, Noble-haeusslein LJ (2009) Traumatic brain injury: an overview of pathobiology with emphasis on military populations. J Cereb Blood Flow Metab 30(2):255–266. https://doi.org/10.1038/jcbfm.2009.203
6. Li Y, Fan H, Gao X (2022) Ballistic helmets: recent advances in materials, protection mechanisms, performance, and head injury mitigation. Compos. Part B 238(April):109890. https://doi.org/10.1016/j.compositesb.2022.109890
7. Yang JJ, Dai J (2014) Simulation-based assessment of rear effect to ballistic helmet impact simulation-based assessment of rear effect to ballistic helmet impact. https://doi.org/10.3722/cadaps.2010.59-73
8. Palta E, Hongbing F, Weggel DC (2018) Finite element analysis of the advanced combat helmet under various ballistic impacts. Int J Impact Eng 112(September):125–143. https://doi.org/10.1016/j.ijimpeng.2017.10.010

Additive Manufacturing and Sustainability: The Mediating Role of Supply Chain

Praveen Kumar Dwivedi, Sumit Chawla, Manjeet, Saurabh Agrawal, Ranjana Mishra, and Mukesh Kumar

Nomenclature

AM Additive Manufacturing

1 Introduction

Emerging Industry 4.0 technologies can affect the performance of manufacturing systems, and the selection of any disruptive technology can be both a strategic opportunity as well a threat for organizations. Hence, the right selection of such technologies would significantly affect the success of the firm. In traditional manufacturing, most of the parts are manufactured either by cutting away sections in raw materials or by molding the material in molten form. Additive Manufacturing (AM) is a novel production technique, which is transforming the whole product life cycle. AM's unique ability to produce parts of complex designs, reduced manufacturing costs (less material waste, less assembly steps due to part consolidation, and no need for tools and fixtures), is contributing to both environmental and economic sustainability. In this era of global competition, attention is increasing toward the environmental and social responsibilities of organizations and purposive and proportionate choice of technology would certainly help to achieve these goals. AM presents huge potential benefits in terms of sustainability perspectives with the mediating role of the supply chain. Industries have been attracted since they could benefit from the

P. K. Dwivedi (✉) · S. Chawla · Manjeet · S. Agrawal · R. Mishra
Department of Applied Science, Bharati Vidyapeeth's College of Engineering, New Delhi 1110063, India
e-mail: praveendwivedi05@gmail.com

M. Kumar
Indian Air Force, New Delhi, India

V. S. Sharma et al. (eds.), *Machining and Additive Manufacturing*, Lecture Notes in Mechanical Engineering, https://doi.org/10.1007/978-981-99-6094-1_16

implementation of AM by accelerating and making cost-effective the new product development processes [1], producing customized products [2] and presenting innovative and complex parts to their customers [3]. On this note, the following key issues related to AM role in supply chain and its opportunities for sustainability can be summarized as follows:

1. What are the applications of AM in the supply chain?
2. What are the key factors that influence the implementation of AM in the supply chain, and how can these factors be addressed to promote sustainability?
3. What is the impact of AM supply chain on Sustainable Business Models?

On this backdrop, it is required to look at the diffusion of this technology and study its important factors in implementing AM technology into firms. Thus, this study provides a conceptual framework laying out the important AM implementation factors, which promote sustainability in the supply chain.

2 Additive Manufacturing and Supply Chain

The right product at a lower cost with a smaller lead time has emerged as a critical success factor for manufacturing companies. In fact, capturing the real need and designing the right products with less utilization of company resources is the key to sustainability in the present competitive era [4, 5]. As a result, product varieties are increasing to satisfy the diversified needs of customers [6, 7]. AM is expected to incline supply chains toward more localized approaches [2], which would fundamentally change supply chain management due to changes in the replacement techniques, packaging and labeling, production costs, and storage. The next decade is likely to see some critical changes in supply chain elements and steps. Looking at the emerging state of the application of AM in the processes of different industries, it is required to study the role of AM in supply chain management which has not yet been dealt with in depth.

3 Additive Manufacturing for Sustainable Business Models

In recent years there has been considerable interest in how AM technologies will shape the business model of manufacturing firms. This section concerns the incremental change in position and earnings in the market followed by disruptive change, which will increase the current distribution of economic value. Moreover, a lifecycle perspective and consideration of new business models are important factors to consider when assessing the sustainability of AM technology and its potential applications. AM does not require dedicated die or tool development. Overall, AM is a highly versatile technology that offers many benefits, including greater design flexibility, faster prototyping, reduced waste, and the ability to create complex geometries that would be difficult or impossible to produce using traditional manufacturing techniques [3, 8]. However, AM technology diffusion to current manufacturing systems is a little disappointing both in terms of economies involved and technology. High raw material cost and its slow deposition rate through AM, boil down total process feasibility.

3.1 Economic Implications of Additive Manufacturing

The first barrier faced by smaller firms is scale economies, and organizations, which operate below efficient scale economies are thrown out or acquired by bigger ones. It is interesting to note that unit cost under AM manufacturing is independent of the scale of production, whereas in conventional manufacturing, it declines with production increases. Nevertheless, there is still scope for smaller firms to compete in the market if there is not much difference in cost. Niche market segments with specialized heterogeneous demands like customization and shorter lead time can be best served by AM technology. In addition, distributed manufacturing is also one of the main attributes of AM manufacturing systems, and it will lower the transportation cost due to closer to customer production. Indeed, AM is not only a disruptive technology [9] but it has potential to flourish new business models as well as supply chains. Notably, quality issues and return policy may discourage some potential customers from buying products produced with AM technology. Since AM is supposed to enable distributed supply chain and manufacturing system, the issue of intellectual property rights is the most severe economic consequence of AM [10]. To conclude, Table 1 summarizes opportunities and limitations from a financial perspective for AM-based consumer-centric business models.

Table 1 Opportunities and limitations from an economic perspective

Sr. no	Opportunities	Limitations
1	Innovation through specific design characteristics and rapid prototyping	High raw material cost
2	Mass customization and mass personalization	Poor economies of scale
3	Distributed manufacturing: ease to explore new markets	Intellectual property infringement
4	Reduction in assembly steps and inventory through part consolidation	Higher skill set and experience required to design products
5	Change of product designs without cost penalty in manufacturing	More responsive to heterogeneous customer demands and seasonal variations
6	Customer co-design possibilities	Lack of quality standards, reproducibility

3.2 Technological Implications of Additive Manufacturing

Bower and Christensen [11] highlighted that, most well-managed companies are consistently ahead of their peer industries in commercializing new technologies and as long as these sustaining technologies address the next-generation performance needs of their customers. Same time these firms are hesitant to commercialize those technologies, which appeal to smaller and emerging market segments. AM is argued to be the most suitable technology for mass customization and functionally better-performing products [12]. Aerospace industries has adopted AM technology to reduce cost during new model developments. Due to one-step manufacturing process, AM can incorporate part consolidation, further reducing manufacturing costs and assembly steps. Additionally, AM has better raw material efficiency and less scrap with respect to traditional manufacturing. Numerous multifaceted dynamics are shaping AM technique, but still, it has several restrictions limiting its successful application. The degree of finish, available printable material, and choice of colors are still limited. To conclude, Table 2 summarizes opportunities and limitations from a technology perspective for AM-based consumer-centric business models.

4 Proposed Additive Manufacturing Supply Chain Framework for Sustainability

The conceptual framework for AM supply chain for sustainability is illustrated in Fig. 1. Framework for sustainability with AM in supply chain is developed with a comprehensive approach by structuring four constructs, i.e., AM strategy, AM supply chain, AM production planning, and organizational change along with external challenges. Factors grouped in these four constructs affect successful AM implementation and help to develop a strategy to counter potential external challenges. The following section discusses framework constructs and external challenges.

Table 2 Opportunities and limitations from a technology perspective

Sr. no	Opportunities	Limitations
1	Flexibility in terms of design, variety, on-demand production, production without tooling, and fabricating a wide range of complex geometries	Lack of design tools, limited reproducibility, limited printable materials
2	High manufacturing flexibility	High throughput time
3	High raw material efficiency	Limited printable materials
4	Digital printing without tools and molds	Product size is limited with 3D printer size
5	Ability to fabricate a wide range of complex geometries	Skilled workforce and high experience required

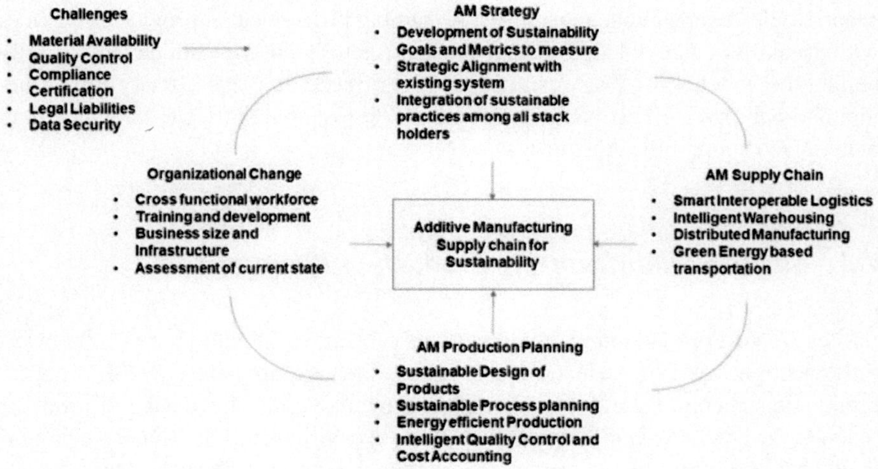

Fig. 1 Additive manufacturing supply chain framework for sustainability

4.1 Additive Manufacturing Strategy

Investing in AM technologies can offer several benefits to businesses, such as faster and more efficient production, increased design flexibility, reduced waste, and lower costs over time. However, it's important to weigh the benefits and risks before making a decision to invest. When it comes to AM, strategic alignment among business, manufacturing, and R&D is important to ensure that the technology is integrated effectively into the overall business strategy. To enhance sustainability, a thorough analysis of the company's operations is required to identify parts or products that can be produced using AM to reduce waste, minimize material usage, and increase resource efficiency. Further integrating sustainable practices, such as using renewable energy sources, reducing water usage, and minimizing waste with production processes, can move a step further toward sustainability. Therefore, it is proposed to develop a detailed plan that outlines the specific sustainability goals and targets the company wants to achieve, the steps required to achieve these goals, and the metrics used to measure progress.

4.2 Organizational Change

To sustain the benefits of AM and achieve ongoing improvements in sustainability, organizations must foster a culture of innovation and continuous improvement. This involves encouraging employees to identify and implement new ideas and processes that further reduce waste, lower energy consumption, and decrease transportation emissions. The successful adoption of AM requires the development of new skills and capabilities. Organizations must invest in training programs that provide employees with the skills and knowledge needed to use the new technology effectively. To fully realize the benefits of AM, organizations must invest in the necessary infrastructure and equipment. This includes acquiring the essential software, hardware, and materials to support the AM process.

4.3 Additive Manufacturing Production Planning

Successful AM production planning requires careful consideration of design, material selection, machine selection, post-processing requirements, quality control, production scheduling, and inventory management. Sustainable design of products refers to the practice of creating products that are environmentally friendly, economically viable, and socially responsible throughout their entire life cycle. This involves considering the impact of a product from its raw material extraction to its eventual disposal or recycling and finding ways to minimize any negative environmental or social effects while maximizing benefits [13]. AM has the ability to create complex, custom-made parts with high precision and efficiency. However, as with any manufacturing process, ensuring the quality of the final product is critical to its overall sustainability. One of the key aspects of quality control in AM is to select from sustainable material options, such as bioplastics or recycled materials, which have a lesser environmental impact. Continuous monitoring through sensors and monitoring software can help to identify and prevent errors that could lead to material waste and increased energy consumption. Eventually, adhering to recognized quality standards and certifications, such as ISO 9001, can help ensure that the AM process is conducted in a sustainable and environmentally responsible manner. Diegel et al. [13] highlighted that AM has the potential to target cost efficiency and product quality while maintaining sustainable product design. However, training to integrate sustainability and certifying new components as sustainable design is still a challenge.

4.4 Additive Manufacturing Supply Chain

Intelligent warehousing and smart, interoperable logistics can work together to promote both economic and environmental sustainability in AM. By combining real-time tracking and intelligent warehousing with smart, interoperable logistics, manufacturers can optimize transportation routes and reduce the carbon footprint associated with transportation. This can improve the environmental sustainability of AM while also reducing transportation costs and improving economic sustainability. Further, by using intelligent warehousing and smart, interoperable logistics, manufacturers can facilitate collaboration between different actors in the AM supply chain, including designers, manufacturers, and logistics providers. This can improve communication, coordination, and efficiency, which can lead to cost savings and improved economic sustainability. By using green energy transportation, such as electric or hybrid vehicles, manufacturers can further reduce the carbon footprint of transportation. This can reduce the environmental impact of AM while also reducing transportation costs, which can improve the economic sustainability of the process.

4.5 Challenges to the Sustainability of Additive Manufacturing Supply Chain

While AM has the potential to promote economic and environmental sustainability, there are still several challenges that need to be addressed to achieve these goals in the AM supply chain. AM can involve multiple suppliers and partners, which can increase the complexity of the supply chain. This can make it difficult to ensure that sustainable practices are being used throughout the process and can increase the risk of waste or environmental impacts. There is still a lack of industry-wide standards for AM, particularly in the areas of quality control and sustainability. AM is still limited in the number and types of materials that can be used. This can limit the ability to use sustainable materials and can result in waste and environmental impacts from the production of non-sustainable materials. The growth of AM has led to an increased focus on regulatory and environmental legislation that governs the AM supply chain. Ensuring compliance with these regulations can be time-consuming and costly, particularly for smaller companies. Many industries, such as aerospace and medical, require certification for components and products. Ensuring that AM products meet certification requirements can be challenging, particularly for new or innovative products. As AM becomes more connected, there is an increased risk of cyberattacks that can compromise the security of sensitive information. Regulations related to data security can add additional costs and complexity to the supply chain.

5 Conclusion and Future Research Directions

Organizations can develop a sustainable additive manufacturing supply chain that reduces waste, lowers energy consumption, and minimizes transportation emissions. This approach can help organizations build a more sustainable and resilient supply chain that can better meet the needs of customers and stakeholders. The conceptual framework for the AM supply chain in the context of sustainability can help businesses reduce their environmental impact and promote sustainable manufacturing practices. By prioritizing sustainability in their AM processes, businesses can contribute to a more sustainable future. Legislation and regulation related to the AM supply chain are still evolving, and it is important for manufacturers and researchers to stay up-to-date with the latest developments. As AM continues to grow in popularity, it is likely that new regulations will be introduced to address emerging concerns related to the technology. In conclusion, additive manufacturing (AM) has the potential to transform the manufacturing industry, including the supply chain, by enabling customization, reducing material waste, and improving energy efficiency. However, to fully realize the potential of AM for sustainability, there are still many challenges that need to be addressed. Future research in sustainable materials, process optimization, supply chain optimization, circular economy, design for AM, and policy and regulation can help further the development of sustainable manufacturing practices through AM and promote the adoption of the technology in industry. By addressing these challenges, the additive manufacturing supply chain can contribute to a more sustainable future. It is important for the industry and researchers to work together to achieve a sustainable future through additive manufacturing.

References

1. Chen Z (2016) Research on the impact of 3D printing on the international supply chain. In: Advances in materials science and engineering
2. Bogers M, Hadar R, Bilberg A (2016) Additive manufacturing for consumer-centric business models: implications for supply chains in consumer goods manufacturing. Technol Forecast Soc Change 102:225–239
3. Ford SLN (2014) Additive manufacturing technology: potential implications for US manufacturing competitiveness. J Int Com Econ 6:40
4. Chao RO, Kavadias S (2008) A theoretical framework for managing the new product development portfolio: when and how to use strategic buckets. Manag Sci 54:907–921. https://doi.org/10.1287/mnsc.1070.0828
5. Griffin A, Page AL (1996) PDMA success measurement project: recommended measures for project development success and failure. J Prod Innov Manag 13:478–496
6. Kamalini R (2003) Managing product variety: an integrative review and Research Di. Prod Oper Manag 12:79–101
7. Pine B II, Victor B (1993) Making mass customization work. Harv Bus Rev 71:108–117
8. Haleem A, Javaid M (2019) Additive manufacturing applications in industry 4.0: a review. J Ind Integr Manag 4:1930001
9. Hannibal M, Knight G (2018) Additive manufacturing and the global factory: disruptive technologies and the location of international business. Int Bus Rev 27:1116–1127
10. Holland M, Stjepandić J, Nigischer C (2018) Intellectual property protection of 3D print supply chain with blockchain technology. In: 2018 IEEE international conference on engineering, technology and innovation (ICE/ITMC). IEEE, pp 1–8
11. Bower JL, Christensen CM (1995) Disruptive technologies: catching the wave
12. Dwivedi PK, Kumar G, Singh RC (2021) Challenges for mass customization in industry 4.0 environment: an analysis using fuzzy TOPSIS approach. In: Recent advances in mechanical engineering: select proceedings of RAME 2020. Springer, pp 1079–1089
13. Diegel O, Singamneni S, Reay S, Withell A (2010) Tools for sustainable product design: additive manufacturing

Leverage of Metal 3D Printing Technology in the Automotive Industry

Altaf Khan, Akant Kumar Singh, and Navdeep Sharma Dugala

Nomenclature

CAD	Computer-Aided Design
MAM	Metal Additive Manufacturing
DED	Directed Energy Deposition
DLF	Directed Light Fabrication
EBM	Electron Beam Melting
PBF	Power Bed Fusion
3D	Three Dimensional

1 Introduction

Additive manufacturing, which is commonly referred to as 3D printing, is a method of producing three-dimensional objects by adding successive layers of material on top of each other. It was first developed in the 1980s and is divided into three main categories based on the type of raw material used in additive manufacturing technology that is solid, liquid, and powder. Figure 1 shows the history and evolution

A. Khan (✉) · N. S. Dugala (✉)
Automobile Engineering Department, Chandigarh University, Gharuan, Punjab 140413, India
e-mail: altafkhankolimkk@gmail.com

N. S. Dugala
e-mail: hod.automobile@cumail.in

A. K. Singh (✉)
Mechanical Engineering Department, Chandigarh University, Gharuan, Punjab 140413, India
e-mail: akant.e13980@cumail.in

V. S. Sharma et al. (eds.), *Machining and Additive Manufacturing*, Lecture Notes in Mechanical Engineering, https://doi.org/10.1007/978-981-99-6094-1_17

of additive manufacturing technology. One of the most popular methods of additive manufacturing is metal additive manufacturing (MAM), which allows for the production of complex, customized metallic parts directly from a 3D computer-aided design (CAD) model. MAM has several advantages over traditional manufacturing techniques, including the ability to produce desired parts on demand and reduced design cycle time. It is also profitable for low and medium-volume production. The versatility of additive manufacturing, or 3D printing, has made it an attractive option for a wide range of industries, including aerospace, automotive, medical, and energy, among others [1].

In recent years, 3D printing has emerged as a key technology in the manufacturing industry. With the ability to produce complex and customized parts from 3D CAD models, this process has revolutionized the way in which products are designed and produced. Instead of the traditional subtractive manufacturing techniques that involve removing material from a part, 3D printing builds up layers of material to create the desired shape. This 3D printing process also eliminates the need for molds, jigs, and fixtures. This toolless approach allows for greater design flexibility and a shorter time to market, while also reducing energy consumption. This 3D printing technology is used in a variety of practices, including rapid prototyping, rapid tooling, composite parts, and direct part manufacturing [2].

In short, metal 3D printing is a technology that is used to produce metal components using a layer-by-layer process. This process typically involves utilizing specialized software, known as computer-aided design (CAD) software, to develop a digital model or design of the specific component. The virtual design is then utilized to direct and oversee the printing process. There are several types of metal 3D printing technologies, including Directed Energy Deposition (DED), Electron Beam Melting (EBM), Binder Jetting, Powder Bed Fusion (PBF), and Sheet Lamination.

2 Metal 3D Printing Technologies

2.1 Directed Energy Deposition (DED)

It is also known as Directed Light Fabrication (DLF), is an additive manufacturing technology that uses a laser or electron beam to fuse metal powder into a solid form. The process involves depositing metal powder on the surface of the build platform, followed by the application of a high-energy beam that melts and fuses the powder into a solid layer. The process is repeated layer by layer until the desired final shape is achieved. DLF is versatile and highly accurate, making it ideal for applications that require tight tolerances and high-quality surface finishes. It is also useful for repairing or restoring damaged parts and can be used to create large parts as the laser beam can be easily repositioned. DLF's speed, efficiency, and versatility make it a valuable tool for manufacturers looking to produce complex parts with precision and

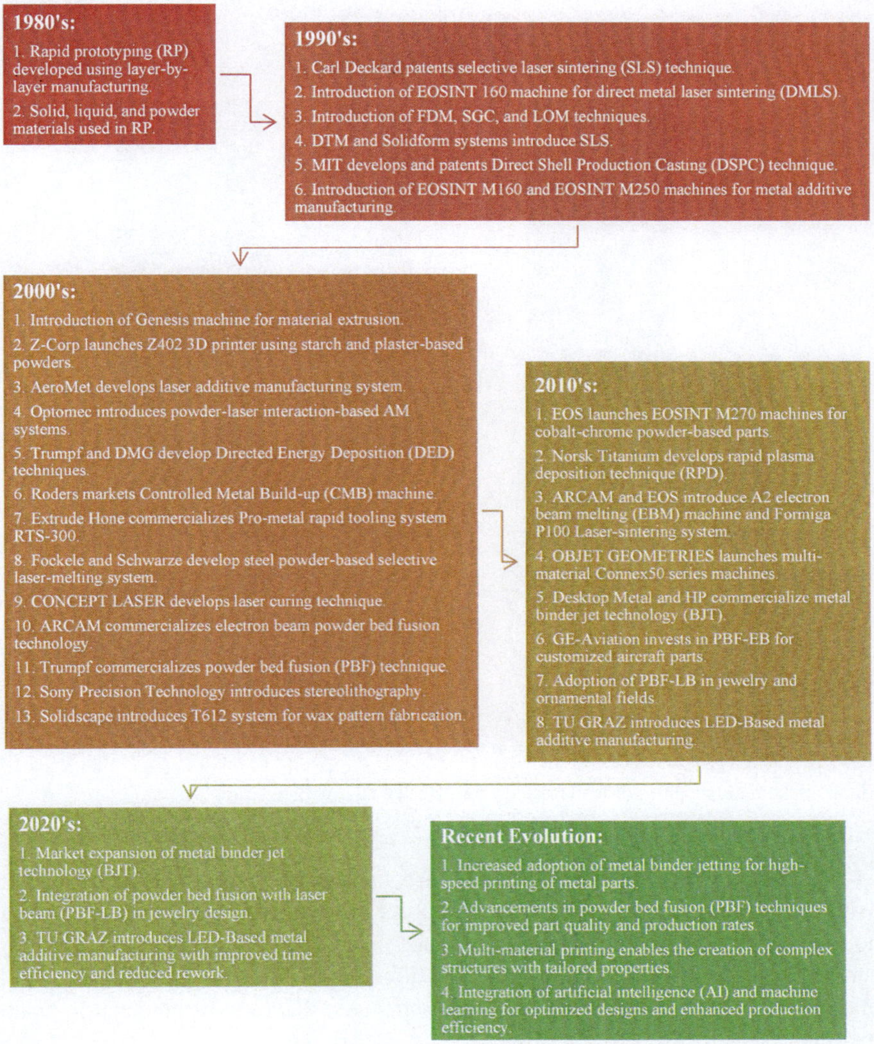

Fig. 1 History and evolution of additive manufacturing processes

efficiency [3–5]. Figure 2 shows the schematic diagram of the laser powder-directed energy deposition.

2.2 Electron Beam Melting (EBM)

It is another type of metal 3D printing technology that works by using high-powered electron beams to melt the metal powder. The process starts by filling a build chamber

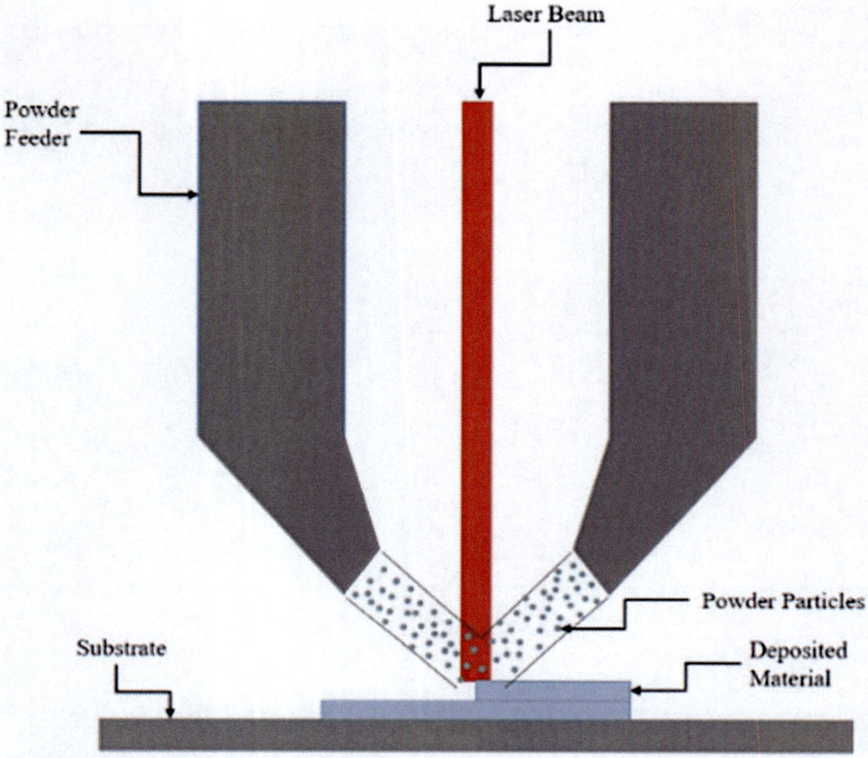

Fig. 2 Schematic diagram of the laser powder directed energy deposition

with a layer of metal powder, which is then melted and deposited layer by layer using an electron beam. The melted material solidifies quickly, producing dense and homogeneous parts with high-quality surface finishes. EBM is commonly used for aerospace, medical, and automotive applications due to its high accuracy, speed, and material strength [6]. Figure 3 shows the schematic diagram of the electron beam melting.

2.3 Binder Jetting

It is a type of metal 3D printing process that works by depositing metal powder and a binding agent layer by layer. The binding agent holds the powder in place while it is being processed, forming the final product. The process is known for its speed, affordability, and ability to produce large parts with high resolution. However, the parts produced by binder jetting are typically not as strong as those produced by other metal 3D printing processes, and they may require further processing to reach

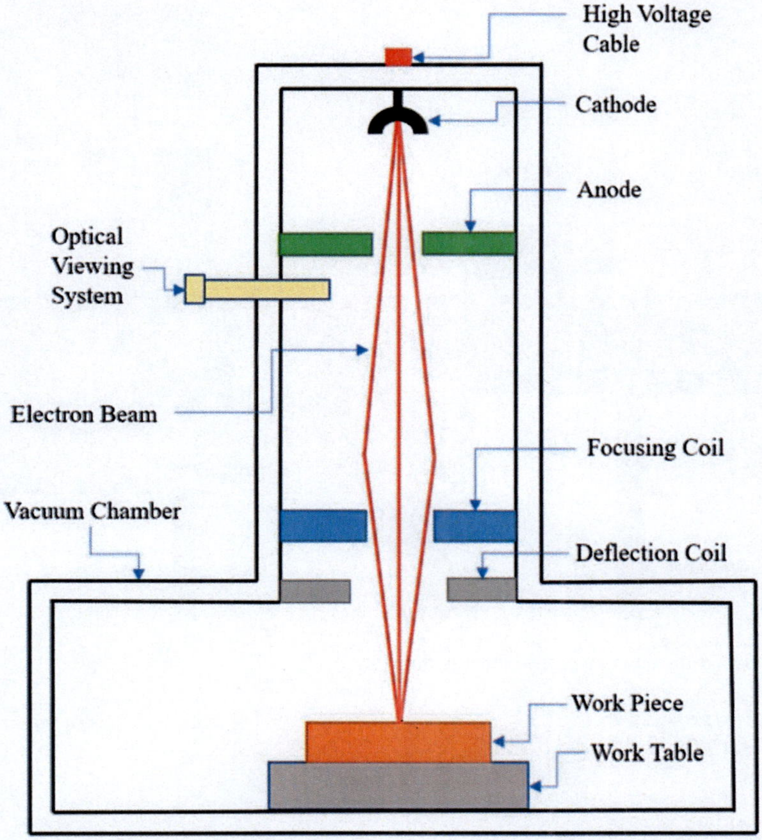

Fig. 3 Schematic diagram of the electron beam melting

the desired strength and durability [5, 7, 8]. Figure 4 shows the schematic diagram of the binder jetting.

2.4 *Powder Bed Fusion (PBF)*

It is a type of metal 3D printing process that works by fusing metal powder into a solid part. The process starts by spreading a layer of metal powder on a build platform and then melting it with a high-powered laser. The laser then moves across the powder, fusing it layer by layer to form the final product. PBF is known for its high precision, speed, and material strength, making it ideal for aerospace, medical, and industrial applications [5, 9, 10]. Figure 5 shows the schematic diagram of the powder bed fusion.

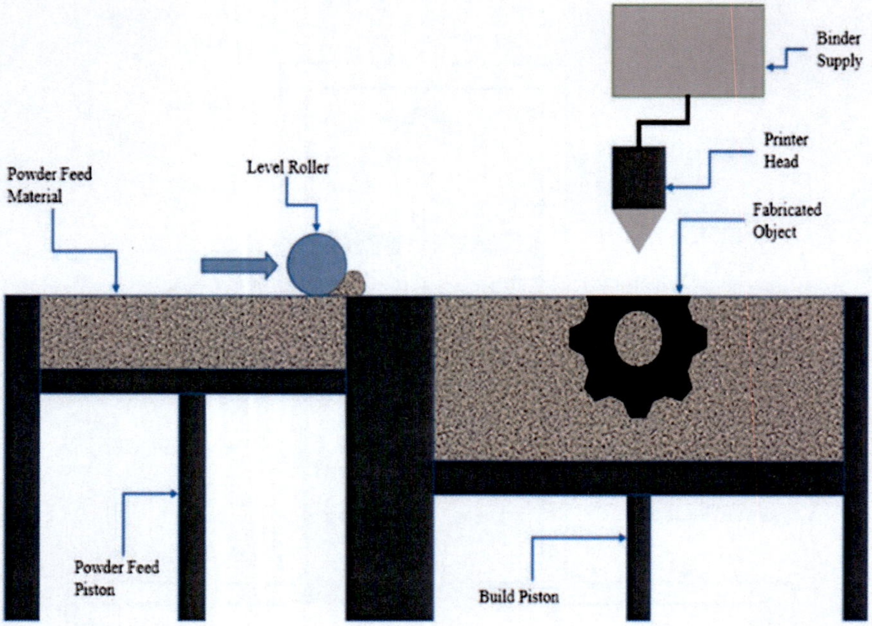

Fig. 4 Schematic diagram of the binder jetting

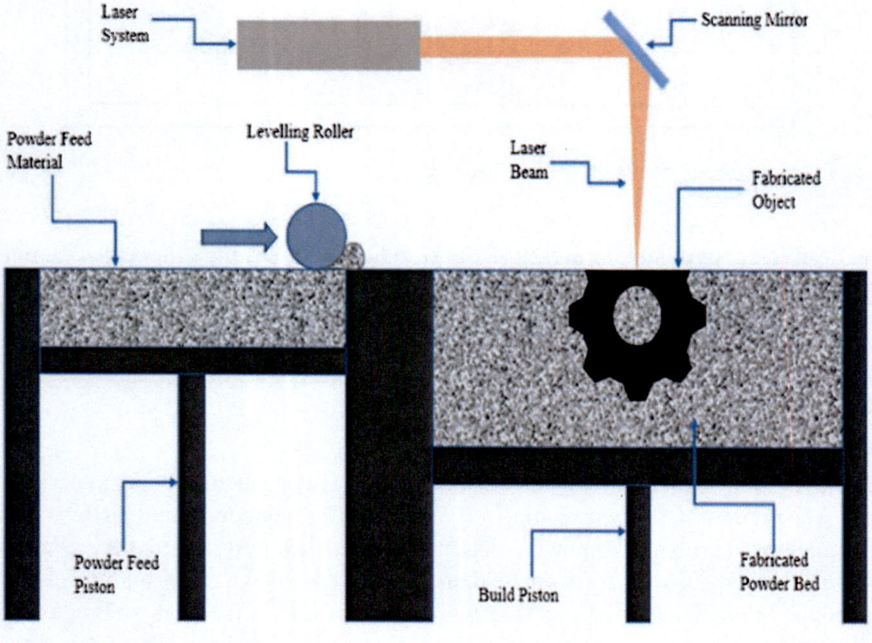

Fig. 5 Schematic diagram of the powder bed fusion

2.5 *Sheet Lamination*

Sheet Lamination is a metal 3D printing technology process that works by layering sheets of metal to form the final product. The process starts by cutting the metal sheets into specific shapes and then layering them on top of each other. The layers are then bonded together using heat and pressure to create the final product. Sheet lamination is a fast and affordable process, making it ideal for producing large parts with high strength and durability. However, the process is not as precise as other metal 3D printing processes, making it better suited for applications that don't require high accuracy [10, 11]. Figure 6 shows the schematic diagram of the sheet lamination.

3 Materials Used in Metal 3D Printing Technology

Metal 3D printing technology can be performed using various types of metal materials, including stainless steel, aluminum, cobalt-chrome, titanium, nickel-based alloys, and copper. The choice of material depends on the application and the properties required, such as strength, stiffness, toughness, and thermal resistance. In the automobile industry, titanium, aluminum, and stainless steel are the most commonly used materials for metal additive manufacturing. Metal 3D printing technology is becoming increasingly important in the automotive industry due to its ability to produce high-performance, lightweight, and cost-effective components, as well as its ability to produce customized and complex parts, reducing lead times and production costs. Here is a brief overview of the most commonly used metals in 3D printing for automobiles and their properties, advantages, and applications [7, 8, 10, 11].

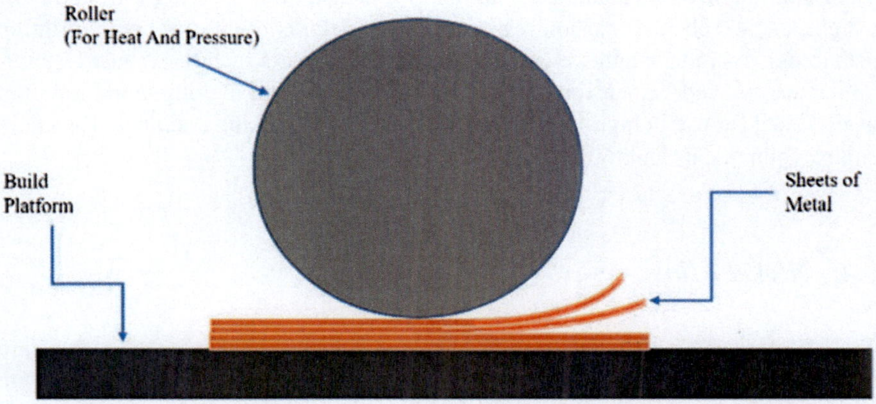

Fig. 6 Schematic diagram of the sheet lamination

3.1 Aluminum

Aluminum is one of the most commonly used metals in 3D printing technology for the automotive industry due to its lightweight, corrosion-resistant, and thermally conductive properties. This metal is known for its high strength-to-weight ratio, which makes it ideal for lightweight applications and for improving fuel efficiency. In addition, aluminum has good machinability, making it easier to produce complex parts in large quantities. Applications of aluminum in the automotive industry include brackets, engine components, heat exchangers, and various body parts such as doors and hoods [7, 10, 11].

3.2 Steel

Steel is another popular metal in the automotive industry, known for its strength, durability, and stiffness. This metal can be used in its traditional form or as a powdered metal for 3D printing technology. Steel is ideal for high-load-bearing parts such as gears, cogs, and other transmission components. In addition, steel is known for its ability to withstand high-stress and high-temperature environments, making it suitable for use in various applications such as engine components, drive shafts, and suspension systems [7, 10, 11].

3.3 Titanium

Titanium is highly in demand in the automotive industry due to its high strength, low weight, and excellent corrosion resistance. This metal is often used in 3D printing technology to produce lightweight, high-performance parts such as exhaust systems, turbochargers, and suspension components. Due to its high melting point, titanium is also ideal for use in high-temperature environments, making it suitable for use in engine components and turbochargers [7, 10, 11].

3.4 Nickel Alloys

Nickel alloys such as Inconel and Hastelloy are generally known for their corrosion resistance, high strength, and ability to withstand high temperatures. These alloys are often used in automotive applications such as turbochargers, exhaust systems, and engine components due to their high performance and durability. These alloys are also known for their excellent resistance to oxidation and high-temperature corrosion, making them ideal for use in high-performance vehicles and racing applications [7, 10, 11].

3.5 *Copper*

Copper is another popular metal in the automotive industry, known for its good thermal and electrical conductivity. This metal is ideal for use in electric vehicle (EV) components, heat exchangers, and electrical wiring due to its high conductivity and thermal resistance. In addition, copper is known for its high strength and toughness, making it ideal for use in high-performance and racing applications. Copper is also used in 3D printing technology to produce high-performance components for racing and high-performance vehicles [7, 10, 11].

3.6 *Cobalt-Chrome*

Cobalt-chrome is a metal that is commonly used in medical and aerospace applications, but it is also gaining popularity in the automotive industry due to its high strength and toughness. This metal is known for its excellent resistance to wear and corrosion, making it ideal for applications that require high durability and reliability. Applications of cobalt-chrome in metal 3D printing technology for the automotive industry include engine components, turbochargers, and heat exchangers [7, 10, 11].

4 Applications of Metal 3D Printing Technology in the Automotive Industry

The automotive industry is one of the industries that have greatly benefited from metal 3D printing. The technology has been used to produce several components, including engine components, suspension parts, exhaust systems, and body parts, among others. Some of the applications of metal additive manufacturing in the automobile industry include [10–12].

4.1 *Engine Components*

The use of metal 3D printing technology in the production of engine components has revolutionized the automobile industry. The ability to produce complex geometries with reduced weight has enabled manufacturers to produce highly efficient engines.

For example, the production of turbochargers using metal 3D printing technology has allowed for the creation of designs with intricate internal passages, which were previously not possible using traditional manufacturing methods. This has resulted in improved engine performance and reduced emissions, as lighter components reduce the overall weight of the vehicle, which in turn improves fuel efficiency [10, 11, 13].

4.2 Suspension Parts

Suspension parts play a vital role in the stability and handling of a vehicle. Metal 3D printing technology has allowed manufacturers to produce suspension components, such as control arms, with improved strength and stiffness, which enhances the overall performance of the vehicle. The ability to produce complex geometries with reduced weight has also resulted in improved fuel efficiency, as lighter components reduce the overall weight of the vehicle. This, in turn, reduces emissions and results in a more environmentally friendly vehicle [10, 11, 13].

4.3 Exhaust Systems

Exhaust systems play a crucial role in reducing emissions and improving fuel efficiency. The use of metal additive manufacturing in the production of exhaust systems has enabled manufacturers to produce parts with complex geometries, which were previously not possible using traditional manufacturing methods. This has resulted in improved performance and reduced emissions, as the ability to produce lighter components results in reduced weight and improved fuel efficiency. In addition, metal 3D printing technology has enabled the production of highly durable components, which have longer life spans and require less maintenance compared to traditional components [10, 11, 13].

4.4 Body Parts

The production of body parts, such as spoilers and grilles, using metal 3D printing technology has revolutionized the automobile industry. The ability to produce intricate designs with reduced weight has enabled manufacturers to produce highly aerodynamic components that not only improve fuel efficiency and reduce emissions but also provide a stylish appearance to the vehicle. The use of metal additive manufacturing has also resulted in improved durability, as components produced using this technology are highly resistant to wear and tear [10, 11, 13].

5 Benefits of Metal 3D Printing Technology in the Automotive Industry

5.1 Customization

Metal 3D printing technology enables manufacturers to produce customized and complex components for different models of vehicles. This results in reduced design and manufacturing time, allowing for more flexibility and quicker time to market [14].

5.2 Lightweight Components

Metal 3D printing technology can produce components that are lighter than traditionally manufactured components, which can reduce the weight of a vehicle and improve fuel efficiency [14].

5.3 Cost Efficiency

Metal 3D printing technology allows for the production of components in small quantities, reducing the cost and waste associated with mass production. This is particularly useful for producing low-volume components or replacement parts [14].

5.4 Complex Geometries

Metal 3D printing technology enables the production of components with complex geometries that are not feasible with traditional manufacturing methods. This allows for new and innovative designs that can improve the performance and efficiency of vehicles [14].

5.5 Environmentally Friendly

Metal 3D printing technology reduces the waste generated in the production process, reducing the environmental impact of manufacturing [14].

6 Challenges of Metal 3D Printing Technology in the Automotive Industry

6.1 High Cost

Metal 3D printing equipment and material costs are still high, making it a more expensive option compared to traditional manufacturing methods [4, 11, 14].

6.2 Lack of Standardization

Metal 3D printing is still in its early stages, and there is a lack of standardization in terms of materials, processes, and equipment. This can lead to inconsistencies in the quality and performance of printed components [11, 14].

6.3 Limited Material Options

The range of materials available for metal 3D printing is still limited, and some materials may not be suitable for use in high-stress environments like automobiles [11, 14].

6.4 Post-Processing Requirements

Metal 3D-printed components often require post-processing to improve their strength and durability. This can be time-consuming and may increase the overall cost of the components [11, 14].

6.5 Quality Control

Maintaining consistent quality control in metal 3D printing technology can be challenging due to the newness of the process and the specialized skills and expertise required [11, 14].

7 Future Scope of Metal 3D Printing in the Automotive Industry

The future of metal 3D printing technology in the automobile industry is quite promising due to the rapid advancement of technology and the decreasing cost of the process. The use of metal 3D printing technology has likely revolutionized the automobile industry by providing a cost-effective and efficient method for the production of high-performance and lightweight components. These components will play a critical role in improving the overall performance and efficiency of vehicles, as they will reduce the weight of the vehicle while maintaining its structural integrity [4, 13, 15].

Moreover, the flexibility offered by metal 3D printing technology allows manufacturers to manufacture components with complex geometries and intricate designs, which will not only improve the aesthetic appeal of vehicles but also provide new design possibilities. This opens up new avenues for innovation in the automobile industry and will help in the development of highly customized vehicles [4, 13, 15].

So, the future of metal additive manufacturing in the automobile industry is bright, and we can expect to see significant advancements in this field in the coming years. The application of this technology can revolutionize the approach to designing, producing, and manufacturing vehicles. This advancement is expected to create numerous prospects for innovation within the automobile industry [4, 13, 15].

8 Conclusions

The automotive industry has been on a constant quest to improve its manufacturing processes and produce better-performing vehicles. Here are some points that are concluded from this review paper.

- Metal 3D printing technology is a game-changer in the automobile industry.
- It has enormous potential due to its ability to produce lightweight, high-performing, and intricate parts.
- Metal 3D printing has several advantages over traditional manufacturing, including faster production time, improved performance, and design versatility.
- However, the technology also faces challenges, such as high costs, limited material options, and a lack of standardization.
- Despite these obstacles, the future of metal additive manufacturing in the automotive industry looks promising, with continued advancements in technology and cost reductions expected.
- Metal 3D Printing is set to revolutionize the automotive industry, and its impact will only grow in the coming years.

References

1. Pratheesh Kumar S, Elangovan S, Mohanraj R, Ramakrishna JR (2021) Review on the evolution and technology of state-of-the-art metal additive manufacturing processes. Mater Today: Proc 46:907–7920. https://doi.org/10.1016/j.matpr.2021.02.567.
2. Gadagi B, Lekurwale R (2020) A review on advances in 3D metal printing. Mater Today: Proc 45:277–283. https://doi.org/10.1016/j.matpr.2020.10.436
3. Gibson I, Rosen D, Stucker B (2015) Directed energy deposition processes. In: Additive manufacturing technologies. Springer New York, pp 245–268. https://doi.org/10.1007/978-1-4939-2113-3_10
4. jue Tang Z et al (2020) A review on in situ monitoring technology for directed energy deposition of metals. Int J Adv Manuf Technol 108, 11–12. Springer, pp 3437–3463. https://doi.org/10.1007/s00170-020-05569-3
5. Dev Singh D, Mahender T, Raji Reddy A (2021) Powder bed fusion process: a brief review. Mater Today: Proc 46:350–355. https://doi.org/10.1016/j.matpr.2020.08.415
6. Kolamroudi MK, Asmael M, Ilkan M, Kordani N (2021) Developments on electron beam melting (EBM) of Ti–6Al–4V: a review. Trans Indian Inst Met 74(4):783–790. https://doi.org/10.1007/s12666-021-02230-9
7. Mostafaei A et al (2021) Binder jet 3D printing—process parameters, materials, properties, modeling, and challenges. Prog Mater Sci 119. Elsevier Ltd. https://doi.org/10.1016/j.pmatsci.2020.100707
8. Ziaee M, Crane NB (2019) Binder jetting: A review of process, materials, and methods. Addit Manuf 28:781–801. Elsevier B.V. https://doi.org/10.1016/j.addma.2019.05.031.
9. Singh R et al (2019) Powder bed fusion process in additive manufacturing: an overview. Mater Today: Proc 26:3058–3070. https://doi.org/10.1016/j.matpr.2020.02.635
10. Shahrubudin N, Lee TC, Ramlan R (2019) An overview on 3D printing technology: technological, materials, and applications. Procedia Manuf 35:1286–1296. https://doi.org/10.1016/j.promfg.2019.06.089
11. Ngo TD, Kashani A, Imbalzano G, Nguyen KTQ, Hui D (2018) Additive manufacturing (3D printing): a review of materials, methods, applications and challenges. Compos Part B: Eng 143: 172–196. Elsevier Ltd. https://doi.org/10.1016/j.compositesb.2018.02.012.
12. Sargini MIM, Masood SH, Palanisamy S, Jayamani E, Kapoor A (2021) Additive manufacturing of an automotive brake pedal by metal fused deposition modelling. Mater Today: Proc 45:4601–4605. https://doi.org/10.1016/j.matpr.2021.01.010
13. Duda T, Raghavan LV (2016) 3D metal printing technology. IFAC-PapersOnLine 49(29):103–110. https://doi.org/10.1016/j.ifacol.2016.11.111
14. Buchanan C, Gardner L (2019) Metal 3D printing in construction: a review of methods, research, applications, opportunities and challenges. Eng Struct 180:332–348. Elsevier Ltd. https://doi.org/10.1016/j.engstruct.2018.11.045
15. Matta AK, Kodali SP, Ivvala J, Jamaleswara Kumar P (2018) Metal prototyping the future of automobile industry: a review, 5. [Online]. www.sciencedirect.comwww.materialstoday.com/proceedings2214-7853

Digital Twins of Hybrid Additive and Subtractive Manufacturing Systems–A Review

Rajat Jain, Nikhil Bharat, and P. Subhash Chandra Bose

Nomenclature

DT	Digital Twin
IOT	Internet of Things
IIOT	Indusrial Internet of Things
HASM	Hybrid Additive and Subtractive Manufacturing Systems

1 Overview of Digital Twins and Hybrid Manufacturing Systems

1.1 Digital Twins for Manufacturing Systems

Digital twins have emerged as a powerful tool for optimizing manufacturing systems. A digital twin is a virtual replica of a physical system that simulates its behavior and performance. By combining sensor data, machine learning, and other technologies, digital twins provide manufacturers with the ability to monitor, analyze, and optimize their manufacturing processes in real time [1]. Digital twins are used across a wide range of industries, including manufacturing, where they have become increasingly popular in recent years. In manufacturing, digital twins enable manufacturers to create virtual replicas of their production processes, which can be used to identify inefficiencies, optimize production, and reduce costs. The potential benefits of digital twins for manufacturing systems are significant, including improved product quality, reduced downtime, increased productivity, and enhanced safety. As technology continues to

R. Jain (✉) · N. Bharat · P. S. C. Bose
Mechanical Engineering Department, National Institute of Technology, Warangal 506004, India
e-mail: rj712074@student.nitw.ac.in

V. S. Sharma et al. (eds.), *Machining and Additive Manufacturing*, Lecture Notes in Mechanical Engineering, https://doi.org/10.1007/978-981-99-6094-1_18

advance, digital twins are poised to play an increasingly important role in the future of manufacturing [2].

1.2 Hybrid Additive and Subtractive Manufacturing Systems

Hybrid additive and subtractive manufacturing systems combine the benefits of both additive and subtractive manufacturing processes, allowing for the production of complex parts with high precision and accuracy. Additive manufacturing involves building up a part layer-by-layer using materials such as plastics, metals, or ceramics, while subtractive manufacturing involves removing material from a block or sheet of material using cutting tools such as drills or mills. By integrating both additive and subtractive processes in a single system, manufacturers can leverage the benefits of each to produce parts with unique geometries and characteristics. Hybrid manufacturing systems also enable manufacturers to produce parts with reduced material waste, faster production times, and lower costs compared to traditional manufacturing methods [3].

There are several different approaches to hybrid additive and subtractive manufacturing, including simultaneous processing, sequential processing, and parallel processing. In simultaneous processing, both additive and subtractive processes are carried out on the same part at the same time, while in sequential processing, the additive and subtractive processes are carried out one after the other. In parallel processing, multiple parts are produced simultaneously, with each part undergoing both additive and subtractive processing steps. Hybrid additive and subtractive manufacturing systems are becoming increasingly popular in a variety of industries, including aerospace, automotive, medical, and consumer goods. The ability to produce complex parts with high precision and accuracy, combined with the reduced material waste and production times, make hybrid manufacturing systems an attractive option for many manufacturers. As the technology continues to advance, hybrid additive and subtractive manufacturing systems are poised to play an increasingly important role in the future of manufacturing, enabling manufacturers to produce parts with greater efficiency, speed, and quality than ever before [4].

1.3 Combination of Digital Twins and Hybrid Manufacturing Systems

The integration of Digital Twins and Hybrid Manufacturing Systems offers a number of benefits to manufacturers. One of the key advantages is the ability to simulate and optimize manufacturing processes before they are implemented in the real world. This can significantly reduce the time and cost involved in product development, as

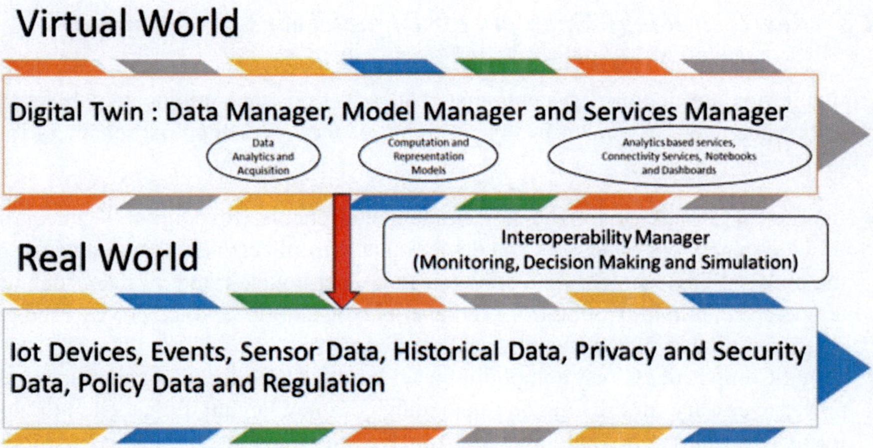

Fig. 1 Conceptual model of the digital twin for hybrid manufacturing systems

well as improve the overall quality of the end product [5]. A conceptual model of DT is shown in Fig. 1.

2 Design and Development of Digital Twins for Hybrid Manufacturing Systems

2.1 A Short Review on Open Source and Commercially Available Digital Twin Solutions

In recent years, there has been a surge of both open source and commercially available Digital Twin solutions. Open-source Digital Twin solutions such as Eclipse Ditto, ThingWorx, and Eclipse Vorto offer developers the freedom to modify and customize the software to meet their specific needs. These solutions provide a strong foundation for building complex and robust Digital Twin systems, with a community of developers constantly contributing to the software's evolution. On the other hand, commercially available Digital Twin solutions such as Microsoft Azure Digital Twins, Siemens Digital Industries Software, ANSYS Twin Builder, and GE Digital provide pre-built Digital Twin models and user interfaces, simplifying the implementation process for companies. These solutions often offer additional features such as advanced analytics, simulation capabilities, and machine learning algorithms, which are critical for predictive maintenance and optimal performance. Overall, both open source and commercially available Digital Twin solutions have their own strengths and limitations. Open-source solutions unhook are ideal for companies looking to develop custom solutions, while commercially available solutions offer a quicker path to implementation and more advanced features [6].

2.2 *Key Technology Enablers for Digital Twins*

Digital Twins are complex systems that require various technological enablers to function effectively as shown in Fig. 2. Some of the key technology enablers for Digital Twins are [7, 8]:

- Internet of Things (IoT): IoT is a fundamental enabler for Digital Twins as it provides a network of sensors and devices that can collect and transmit data from the physical environment. IoT devices can collect a wide range of data, such as temperature, humidity, vibration, and energy consumption, which can be used to create a digital replica of the physical environment.
- Cloud Computing: Cloud computing provides the scalability and computational power necessary for creating and managing complex Digital Twins. Cloud platforms like Microsoft Azure and Amazon Web Services offer advanced services such as data storage, processing, and analytics that are crucial for Digital Twins.
- Artificial Intelligence (AI): AI is used to analyze and process data collected from IoT devices to create insights and predictions for the physical environment. Machine learning algorithms can be trained to detect anomalies and patterns in the data and identify potential issues in the physical system, allowing for proactive maintenance and optimization.
- 3D Modeling and Visualization: 3D modeling and visualization technologies are used to create a digital replica of the physical environment, allowing for better visualization and analysis of the system. Technologies such as Augmented Reality (AR) and Virtual Reality (VR) can also be used to provide a more immersive experience and enable better training and simulation capabilities.
- Data Analytics: Advanced analytics tools are required to process and analyze the vast amounts of data collected from the physical environment. Tools such as data visualization, predictive analytics, and machine learning can be used to gain insights into the system, enabling better decision-making and optimization.
- Cybersecurity: Digital Twins require robust security measures to protect sensitive data and ensure the physical environment is not vulnerable to cyber-attacks.

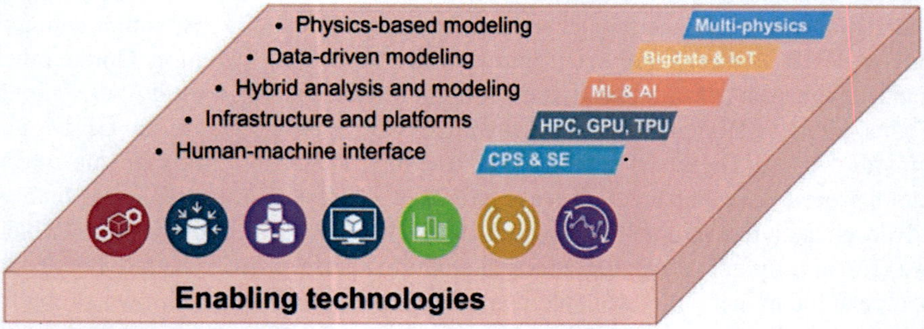

Fig. 2 Key technology enablers for the digital twins [5]

Technologies such as encryption, access control, and intrusion detection systems are crucial to maintain the security of the system.

2.3 Designing and Developing Digital Twins Using Microsoft Azure Digital Twin Platforms

Designing and developing Digital Twins using Microsoft Azure Digital Twin Platforms8 can be achieved through the following steps [5, 6]:

- Identify the physical system and its components that will be modeled in the Digital Twin.
- Define the interface and protocols to collect data from the physical environment, using IoT devices or other sensors.
- Design the Digital Twin model and define the relationships between the components and their behavior.
- Develop the Digital Twin using the Azure Digital Twins platform, using Azure Digital Twins Explorer for modeling and defining the Digital Twin components.
- Define the data schema, including data types, units of measure, and data relationships.
- Develop the data ingestion process, using Azure IoT Hub to collect data from sensors and devices and send them to the Digital Twin.
- Use Azure Stream Analytics to process real-time data and generate insights and predictions about the physical system.
- Use Azure Functions to create workflows that automate actions based on the insights generated from the Digital Twin, such as triggering alerts or notifications.
- Test and validate the Digital Twin by simulating different scenarios, using tools such as Azure Digital Twins Explorer and Power BI for visualization and analysis.

Deploy the Digital Twin to production and monitor its performance, using Azure Application Insights and other monitoring tools to identify and address issues in real time.

2.4 Hybrid Additive and Subtractive Manufacturing System Digital Twin Example

Suppose a manufacturing company produces parts using both additive and subtractive manufacturing processes, with a CNC machine performing the subtractive manufacturing and a WAAM (Wire Arc Additive Manufacturing System) performing the additive manufacturing. The company can create a Digital Twin of this manufacturing system, which would include a digital model of the manufacturing process, the machines, and the parts being produced [9].

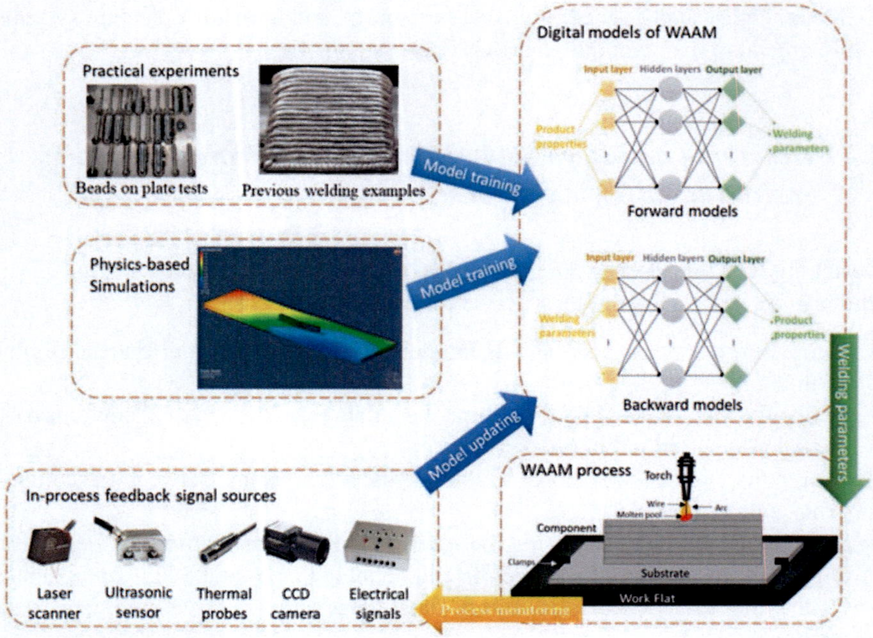

Fig. 3 WAAM–digital twin framework [9]

The Digital Twin (Fig. 3) can be used to simulate the manufacturing process and optimize it for efficiency, quality, and productivity. For example, the Digital Twin can be used to simulate the path of the CNC machine and WAAM, ensuring that they do not interfere with each other during the manufacturing process. It can also be used to optimize the process parameters, such as the speed of the CNC machine and the temperature of the WAAM, to ensure that the parts are produced with the desired quality. Following are the points that explain how DT of WAAM can be used:

- Design: The first step in the WAAM process is to create a 3D model of the part to be printed using computer-aided design (CAD) software. This 3D model can then be imported into a Digital Twin platform, such as Microsoft Azure Digital Twins.
- Simulation: Once the 3D model is imported into the Digital Twin platform, a simulation can be run to analyze the WAAM process. The simulation considers factors such as the properties of the metal wire, the arc voltage, the wire feed rate, and the speed of the print head. The simulation generates data that can be used to optimize the process, such as the optimal parameters for the arc voltage and wire feed rate.
- Printing: Once the simulation is complete, the Digital Twin can be used to control the WAAM machine. The Digital Twin sends commands to the machine, such as the arc voltage and wire feed rate, to ensure that the part is printed according to the optimized parameters generated by the simulation.

- Quality control: As the part is being printed, sensors can be used to collect data on the temperature, the geometry, and the quality of the part. This data is sent back to the Digital Twin platform in real time, where it is analyzed to ensure that the part is being printed correctly. Any deviations from the expected quality can be flagged, and the process can be adjusted accordingly.
- Post-processing: Once the part is printed, it may need to undergo post-processing, such as sanding or polishing, to achieve the desired finish. The Digital Twin can be used to simulate the post-processing steps and optimize the process to achieve the desired finish.

Additionally, the Digital Twin can be used to monitor the performance of the manufacturing system in real time. It can collect data from sensors on the machines and use it to predict when maintenance is needed or to detect any anomalies in the manufacturing process. This can help to reduce downtime and improve overall efficiency (Fig. 3).

2.5 Optimizing the Process Parameters Using Artificial Intelligence Along with Using Real-Time Data Analytics (Microsoft Power Bi) for Condition Monitoring

The following steps can be used for improving the process parameters of WAAM-DT (Fig. 4) [3, 4]:

- Data Preprocessing: The collected data from the sensors needs to be pre-processed to ensure that it is of good quality, consistent, and relevant. This involves data cleaning, data normalization, and data transformation.
- Feature Selection: Feature selection is the process of selecting the most relevant features that are needed to control the WAAM process. This involves selecting the most important parameters such as voltage, current, temperature, and feed.
- Machine Learning Model Development: Machine Learning (ML) models can be developed using tools such as Azure Machine Learning Studio. The ML models can be used to analyze the preprocessed data and predict the output parameters such as the quality of the part.
- AI-Based Control System: An AI-based control system can be developed using machine learning models. This system can be used to control the WAAM process parameters such as voltage, current, temperature, and speed in real time.
- Condition Monitoring using Power BI: Power BI can be used to create a real-time dashboard to monitor the condition of the WAAM process. The dashboard can display the performance of the process, quality of the part, and any anomalies that may occur during the process.
- Real-time Alerts and Notifications: Real-time alerts and notifications can be set up using Power BI to alert the operators and engineers if the WAAM process

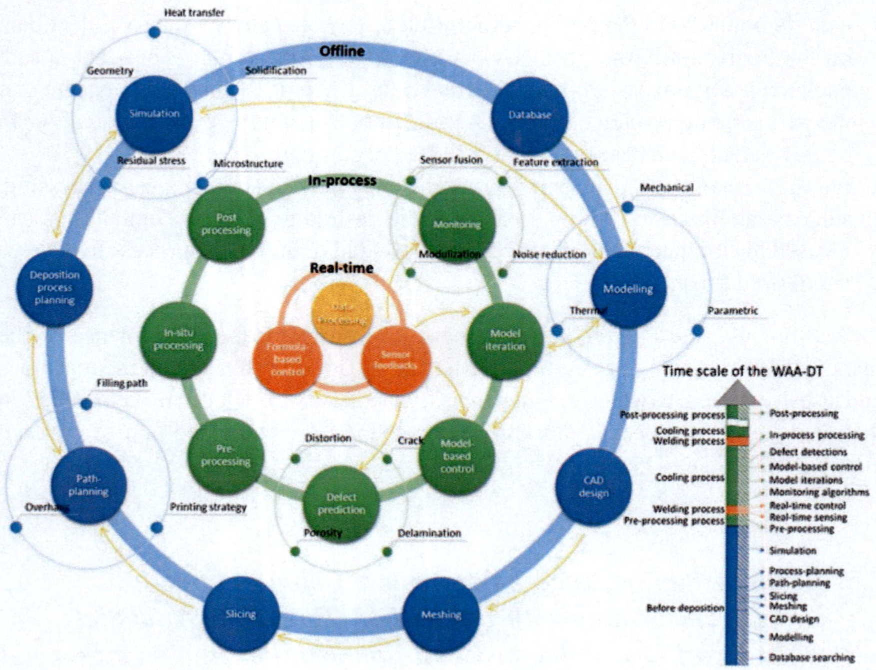

Fig. 4 Real time and offline data monitoring of WAAM-DT for condition monitoring purposes [1, 9, 10]

parameters go out of range, or if there are any anomalies or faults detected during the process.

3 Challenges in the Implementation of Digital Twins for Hybrid Additive and Subtractive Manufacturing Systems

The implementation of Digital Twins for Hybrid Additive and Subtractive Manufacturing (HASM) presents some challenges that need to be addressed to maximize their effectiveness. Here are some of the main challenges [10–15]:

- **Integration of different manufacturing processes**: Digital Twins for HASM need to simulate multiple manufacturing processes, which can be challenging as each process has its own variables and parameters. Integrating the data from different processes and creating an accurate simulation model is crucial for the effective implementation of Digital Twins for HASM.

- **Multi-objective optimization**: Hybrid manufacturing systems have the potential to produce parts with superior performance and functionality compared to single-process manufacturing systems. However, optimizing the manufacturing process for multiple objectives such as quality, speed, and cost can be difficult. The challenge is to develop Digital Twins that can optimize the manufacturing process for these multiple objectives simultaneously.
- **Big data and cybersecurity**: Digital Twins for HASM require a large amount of data from different sources, such as sensor data, material properties, and design parameters. Collecting and managing this data can be a challenge, as even small inaccuracies can have a significant impact on the simulation results.
- **Complexity of the simulation models and high amount of computation resources requirement**: The simulation models for HASM are complex and require high computing power. The challenge is to create accurate and efficient models that can handle the complexity of the hybrid manufacturing process, including the effects of material properties, heat transfer, and fluid flow.
- **Lack of standards and frameworks for Digital twin implementation**: As with Digital Twins for AM, there are no standardized methods for developing Digital Twins for HASM, which can make it difficult to compare and validate results across different systems. This lack of standards can also hinder the adoption of Digital Twins by manufacturers.
- **Requirement of high capital investment**: Building a Digital Twin for HASM requires significant resources, including hardware, software, and personnel. The challenge is to balance the cost of developing and maintaining the Digital Twin with the potential benefits that it can provide.
- **Issues relating to communication networks**: Building quicker and more effective communication interfaces, like 5G is necessary. The use of 5G technology for smart manufacturing is urgently needed due to its numerous benefits, including the expansion of the number of sensors and devices that can be connected, high-speed ubiquitous connectivity, improved reliability and redundancy, and ultra-low power consumption. It is also very important to enable real-time data connectivity and operational efficiency for the Digital Twins.

4 Future Scope of Digital Twins for Hybrid Additive and Subtractive Systems

The future scope of Digital Twins for Hybrid Additive and Subtractive Manufacturing (HASM) is significant, as the technology has the potential to transform the manufacturing industry. Here are some potential future developments and applications of Digital Twins for HASM [13, 14, 16].

- **Advanced process control and optimization**: Digital Twins can be used to optimize the manufacturing process for multiple objectives, such as quality, speed, and cost. In the future, we can expect to see even more advanced optimization algorithms that can adapt to changing process conditions and make real-time adjustments to the manufacturing process.

- **Predictive maintenance**: Digital Twins can be used to monitor the health of machines and predict when maintenance is needed. In the future, we can expect to see more advanced predictive maintenance algorithms that can detect potential issues before they occur and recommend preventative measures.
- **Supply chain optimization**: Digital Twins can be used to optimize the supply chain for HASM, including material sourcing, transportation, and inventory management. In the future, we can expect to see more integrated supply chain models that use Digital Twins to optimize the entire manufacturing process.
- **Adoption of Digital Twins by small and medium-sized enterprises (SMEs)**: Currently, the adoption of Digital Twins is primarily limited to large companies with significant resources. In the future, we can expect to see more SMEs adopting Digital Twins for HASM as the technology becomes more accessible and cost-effective.

Overall, the future scope of Digital Twins for HASM is vast and exciting, with the potential to transform the manufacturing industry and create significant value for manufacturers.

References

1. Gopal L, Singh H, Mounica P, Mohankumar N, Challa NP, Jayaraman P (2023) Digital twin and IOT technology for secure manufacturing systems. Meas: Sens 25. https://doi.org/10.1016/j.measen.2022.100661
2. Bhattacharya M, Penica M, O'Connell E, Southern M, Hayes M (2023) Human-in-loop: a review of smart manufacturing deployments. Systems 11:35. https://doi.org/10.3390/systems11010035
3. Isaja M, Nguyen P, Goknil A, Sen S, Husom EJ, Tverdal S, Anand A, Jiang Y, Pedersen KJ, Myrseth P, Stang J, Niavis H, Pfeifhofer S, Lamplmair P (2023) A blockchain-based framework for trusted quality data sharing towards zero-defect manufacturing. Comput Ind 146. https://doi.org/10.1016/j.compind.2023.103853
4. Nath SV, van Schalkwyk P, Isaacs D (2021) Building industrial digital twins design, develop, and deploy digital twin solutions for real-world industries using Azure Digital Twins. Packt Publishing, Limited
5. Bamunuarachchi D, Georgakopoulos D, Banerjee A, Jayaraman PP (2021) Digital twins supporting efficient digital industrial transformation. Sensors 21. https://doi.org/10.3390/s21206829
6. Damjanovic-Behrendt V, Behrendt W (2019) An open source approach to the design and implementation of digital twins for smart manufacturing. Int J Comput Integr Manuf 32:366–384. https://doi.org/10.1080/0951192X.2019.1599436
7. IEEE Communications Society. Internet of Things, AHSNTC, IEEE Internet of Things (Initiative), Institute of Electrical and Electronics Engineers: GIoTS, Global IoT Summit: 2020 conference proceedings
8. Makarov VL, Bakhtizin AR, Beklaryan GL (2019) Developing digital twins for production enterprises. Bus Inform 13:7–16. https://doi.org/10.17323/1998-0663.2019.4.7.16

9. Mu H, He F, Yuan L, Commins P, Wang H, Pan Z (2023) Toward a smart wire arc additive manufacturing system: a review on current developments and a framework of digital twin. J Manuf Syst 67:174–189. https://doi.org/10.1016/j.jmsy.2023.01.012
10. Botín-Sanabria DM, Mihaita S, Peimbert-García RE, Ramírez-Moreno MA, Ramírez-Mendoza RA, de Lozoya-Santos JJ (2022) Digital twin technology challenges and applications: a comprehensive review
11. Phua A, Davies CHJ, Delaney GW (2022) A digital twin hierarchy for metal additive manufacturing
12. Rasheed A, San O, Kvamsdal T (2019) Digital twin: values, challenges and enablers
13. Xie X, Merino J, Moretti N, Pauwels P, Chang JY, Parlikad A (2023) Digital twin enabled fault detection and diagnosis process for building HVAC systems. Autom Constr 146. https://doi.org/10.1016/j.autcon.2022.104695
14. Kritzinger W, Karner M, Traar G, Henjes J, Sihn W (2018) Digital twin in manufacturing: a categorical literature review and classification. In: IFAC-PapersOnLine. Elsevier B.V., pp 1016–1022
15. Cai Y, Wang Y, Burnett M (2020) Using augmented reality to build digital twin for reconfigurable additive manufacturing system. J Manuf Syst 56:598–604. https://doi.org/10.1016/j.jmsy.2020.04.005
16. Stojanovic N, Milenovic D (2019) Data-driven digital twin approach for process optimization: an industry use case. In: Proceedings-2018 IEEE International conference on big data, big data 2018. Institute of Electrical and Electronics Engineers Inc., pp 4202–4211